GW00382815

ULTIMATE ATLAS
OF THE
WORLD

Philip Steele • Keith Lye

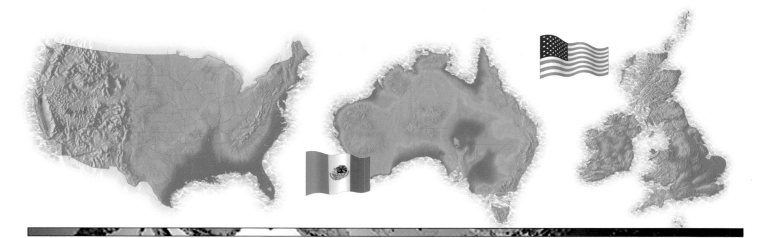

ULTIMATE ATLAS
OF THE
WORLD

Philip Steele • Keith Lye

PARRAGON

Authors
Keith Lye
&
Philip Steele

Editor
Clare Oliver

Designer
Sally Boothroyd

Project Management
Kate Miles

Artwork Commissioning
Susanne Grant & Lynne French

Picture Research
Janice Bracken, Lesley Cartlidge, Graham King, Liberty Mella

Production
Jenni Cozens & Ian Paulyn

Design Director
Clare Sleven

Cartography
Digital Wisdom

Reproduction House
DPI, Saffron Walden

First published in 1999 by Parragon
Parragon
Queen Street House
4 Queen Street
Bath
BA1 1HE, UK

24681097531

Copyright © Parragon 1999

Produced by Miles Kelly Publishing Ltd
Bardfield Centre, Great Bardfield, Essex CM7 4SL

All rights reserved. No part of this publication may be reproduced, stored in a retrieval
system, or transmitted by any means, electronic, mechanical, photocopying,
recording or otherwise, without the prior permission of the copyright holder.

British Library Cataloguing-in-Publication Data
A catalogue record for this book is available from the British Library

ISBN 0-75253-355-5

Printed in Germany

Contents

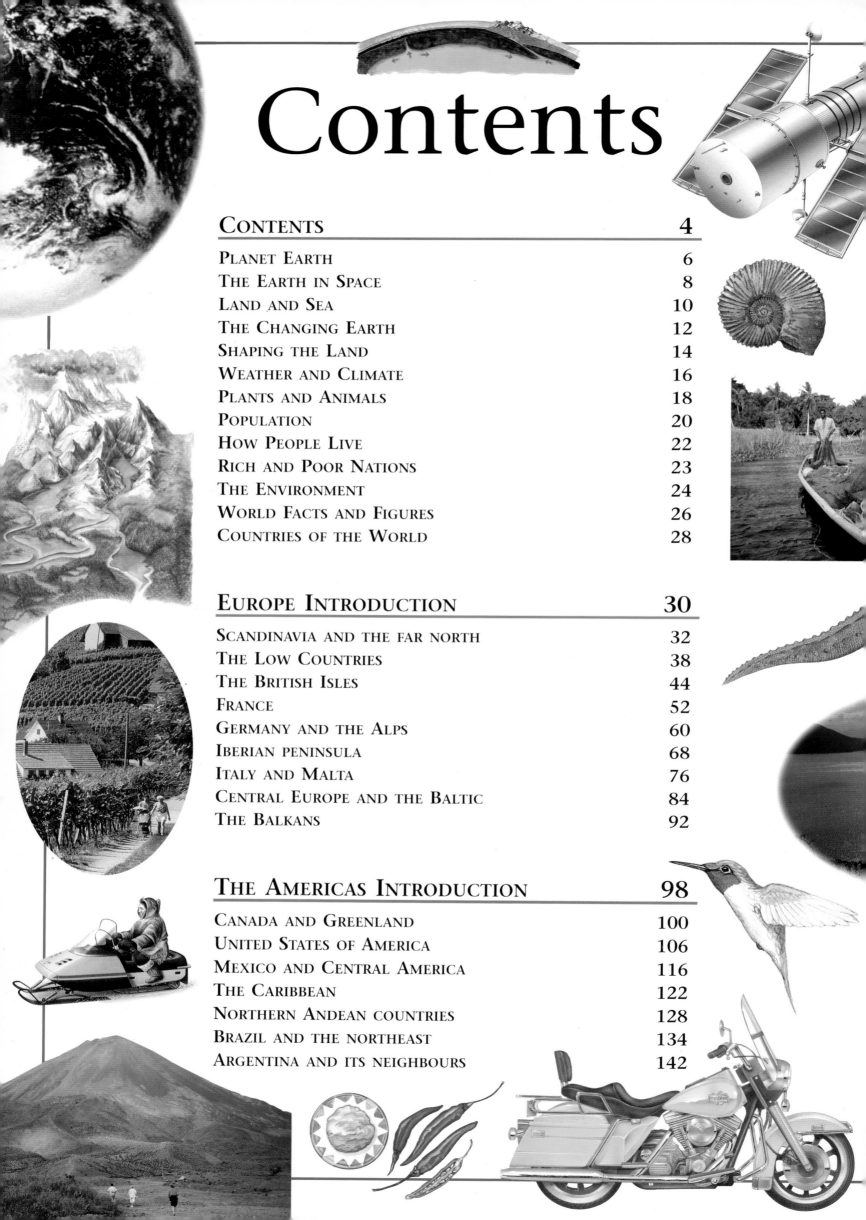

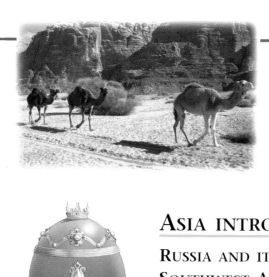

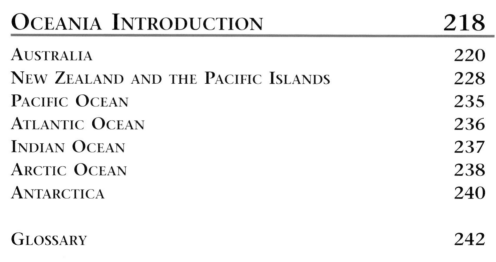

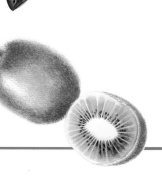

Introduction
Planet Earth

The Earth is a huge sphere of rock, surrounded by air and partly covered by water. The North Pole is the point at the top of the sphere, while the South Pole is at the bottom. The imaginary line connecting the North Pole, the centre of the Earth and the South Pole is called the Earth's axis. It is tilted by about 23.5 degrees.

Another imaginary line runs around the Earth exactly halfway between the two poles. This line is called the Equator. It divides the world into two halves, called hemispheres. The Equator appears on globes, together with other lines that run around the globe parallel to the Equator. These are called parallels, or lines of latitude. Latitude is measured in degrees between the Equator (0 degrees latitude) and the poles (90 degrees latitude). The latitude of any place between them is the angle formed at the centre of the Earth between the place and the Equator.

Other lines on the globe run at right angles to the parallels. These are called meridians, or lines of longitude. The prime meridian (0 degrees longitude) runs through the Greenwich, in London. Other lines of longitude are measured 180 degrees east and 180 degrees west of the prime meridian. On the opposite side of the world from the prime meridian, around 180 degrees east or west, lies the International Date Line.

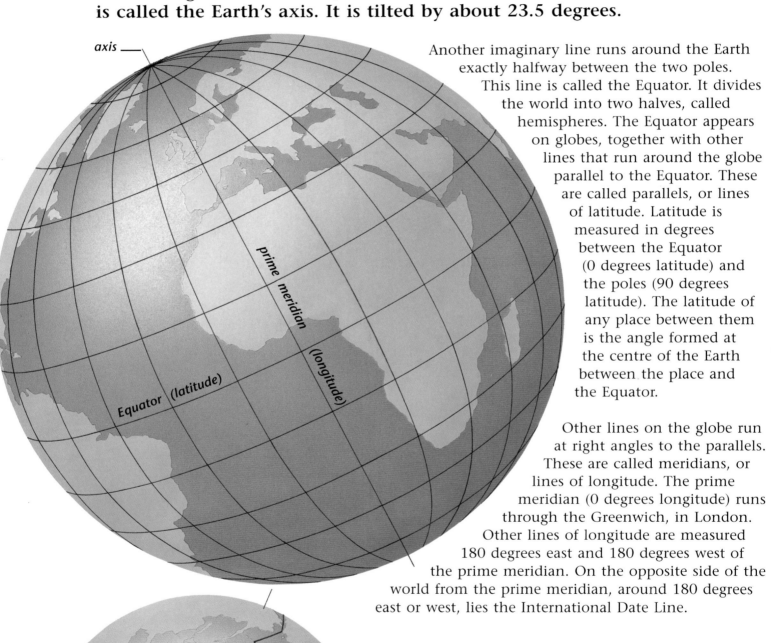

axis

prime meridian (longitude)

Equator (latitude)

Tropic of Cancer

Equator

Tropic of Capricorn

International Date line

Special Lines of Latitude

The hottest part of the world, called the tropics, lies between two special lines of latitude: the Tropic of Cancer, which is 23.5 degrees north and the Tropic of Capricorn, which is 23.5 degrees south. The coldest parts of the world lie north of the Arctic Circle, which is 66.5 degrees north, and south of the Antarctic Circle, which is 66.5 degrees south.

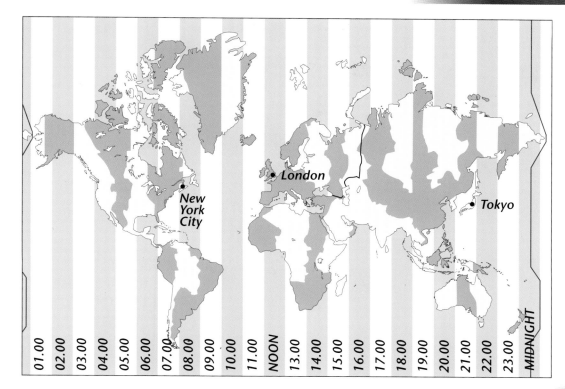

◄ Time zones
Because the Earth rotates once every 24 hours, clocks in various parts of the world show different times. Time zones are measured east and west of the prime meridian. Canada and the United States each have six standard time zones.

▼ Mapping the land
Land surveyors measure the positions and heights of places on the Earth's surface. Their measurements are used to make maps.

To map any area, people called surveyors measure the exact latitude and longitude of a network of points on the Earth's surface. Then they measure the positions of all the land features between the points. With this information, they draw maps of the area. Drawing maps of large areas is made difficult by the fact that the Earth's surface is curved. World maps, which show the entire surface of the globe on a flat piece of paper, must be distorted to ensure that certain things, such as correct areas, distances and directions, are preserved. The only true world map is the globe, because it has a curved surface.

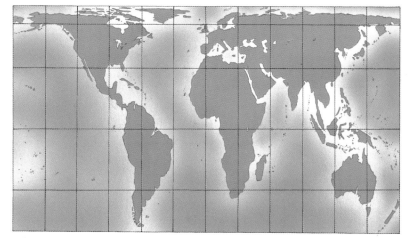

◄ Finding the way
Explorers use compasses to work out directions. Compass needles point towards the magnetic North Pole.

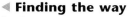

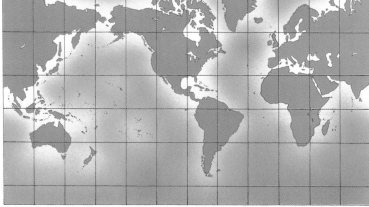

▲ Peter's projection
Map projections are ways of showing the Earth's curved surface on a flat piece of paper. The Peter's projection shows the areas of land masses accurately, but it distorts their shapes.

▲ Mercator's projection
No map projection is completely accurate. Gerardus Mercator's 16th-century projection shows directions accurately and ships' navigators used it to find their way. But this projection distorts areas.

The Earth in Space

The Earth is a tiny speck in space. It is the fifth-largest of the nine planets that rotate around the Sun in the Solar System. The Sun is one of millions of stars that make up the Milky Way galaxy, which, in turn, is one of the billions of galaxies that form the Universe.

1	Sun
2	Mercury
3	Venus
4	Earth
5	Mars
6	Jupiter
7	Saturn
8	Uranus
9	Neptune
10	Pluto

▶ **The Solar System**
Nine planets and their moons rotate around the Sun. Together with other bodies, such as asteroids, comets and meteors, they make up the Solar System.

Equinoxes and Solstices

On two days every year, on March 20th or 21st, and again on September 22nd or 23rd, the Sun is overhead at noon at the Equator. These two days are called equinoxes, a term meaning 'equal night'. This is because everywhere on our planet has 12 hours of darkness and 12 hours of daylight.

However, after March 21st, the northern hemisphere leans increasingly towards the Sun. On June 20th or 21st, the Sun is overhead at noon at the Tropic of Cancer. This day is called the summer solstice in the northern hemisphere and the winter solstice in the southern hemisphere.

After September 23rd, the southern hemisphere leans increasingly towards the Sun until, on December 21st or 22nd, the Sun is overhead at the Tropic of Capricorn. This is the summer solstice in the southern hemisphere and the winter solstice in the northern hemisphere.

▼ **The four seasons**
Places in the middle latitudes experience four seasons in the year as the Earth rotates once around the Sun.

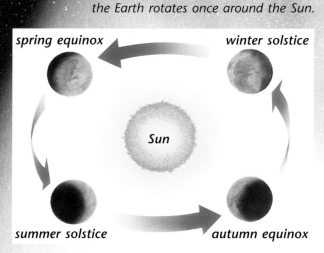

spring equinox   winter solstice

Sun

summer solstice   autumn equinox

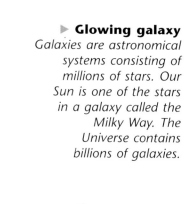

> **Glowing galaxy**
> *Galaxies are astronomical systems consisting of millions of stars. Our Sun is one of the stars in a galaxy called the Milky Way. The Universe contains billions of galaxies.*

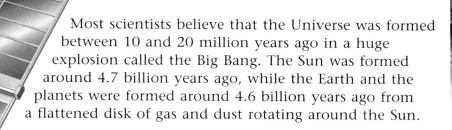

Most scientists believe that the Universe was formed between 10 and 20 million years ago in a huge explosion called the Big Bang. The Sun was formed around 4.7 billion years ago, while the Earth and the planets were formed around 4.6 billion years ago from a flattened disk of gas and dust rotating around the Sun.

Our planet is always on the move. It rotates on its axis once every 24 hours, giving us night and day. It also orbits around the Sun once every 365 days, 5 hours, 48 minutes and 46 seconds. This is called the solar year. Our calendar includes some leap years of 366 days to allow for the difference between the solar and calendar year.

> ▲ **Hubble telescope**
> *Launched into space in 1990, the Hubble telescope has sent back amazing pictures. These show distant galaxies that could not be seen through telescopes on Earth.*

> ▶ **Day and night**
> *The Earth spins on its axis, taking 24 hours, or one day, to complete one revolution. The term day is also used for the period when the Sun is shining on our part of the Earth. But the night, when it is dark because our part of the Earth faces away from the Sun, is also part of the whole 24-hour day.*

Sun

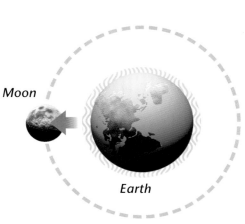

Moon

Earth

Neap tides occur when the Moon and Sun form a right angle with the Earth.

Tides

The gravitational pull of the Moon and, to a lesser extent the Sun, causes bulges in the waters of the oceans. These bulges cause tides. Two high tides and two low tides occur every 24 hours and 50 minutes. The highest tides are called spring tides. These occur when the Moon, Earth and Sun are in a straight line so the pull of the Moon and Sun is combined.

The smallest tidal range (the difference between the levels of high and low tides) occurs when the Moon, Earth and Sun form a right angle. The gravitational pull of the Moon and Sun are then opposed, causing neap tides. Spring and neap tides occur about twice every month.

Introduction
Land and Sea

Land covers about 29.1 percent of the Earth's surface. Most of the land is grouped into seven continents. Water covers nearly 71 percent of the Earth's surface. The largest of the four oceans is the Pacific, which covers about a third of the Earth's surface. People sometimes describe the waters around Antarctica as a fifth ocean, called the Southern or Antarctic Ocean. However, most geographers regard these waters as the southern parts of the Atlantic, Indian and Pacific oceans.

Around most of the continents are shallow seas, covering areas called continental shelves. The continental shelves are flooded parts of the continents. The gently-sloping continental shelves end at the continental slope, which descends steeply to the abyss.

Moving Seawater

Seawater is always on the move. Waves, which make the water move up and down, are caused by the wind. Wind-driven currents carry warm water from the tropics towards the cold, polar regions, while cold currents flow back towards the Equator. The Gulf Stream is a warm current which begins in the Gulf of Mexico. It runs up the coast of the eastern United States and then crosses the Atlantic towards Europe.

▼ **The ocean bed**
Running through the Atlantic Ocean from north to south is a huge, mostly submerged mountain range called the Mid-Atlantic Ocean Ridge.

continent

continental shelf

continental slope

abyssal plain

ocean ridge

oceanic crust

magma

North Atlantic Drift

Arctic Circle

Labrador Current

Oyashio Current

North Pacific Current

Gulf Stream

Canaries Current

Tropic of Cancer

Kuroshio Current

North Equatorial Current

Equator

Monsoon Drift

Australian Current

Peru Current

Tropic of Capricorn

Brazil Current

West Australian Current

Antarctic Circumpolar Current

Antarctic Circle

Coral Islands

Oceanic islands rise from the deep ocean bed. These islands are active or extinct volcanic mountains. The tops of many ancient volcanoes are capped by coral. Coral is a hard rock built up by tiny creatures called coral polyps in warm, shallow seas. By contrast, continental islands, such as the British Isles, are parts of the continental shelf which lie above sea level.

The abyss contains vast muddy plains and huge volcanic mountains, some of which rise from the ocean floor and emerge as islands. Other features are ocean ridges, which are long, underwater mountain ranges. In the middle of these ridges are rift valleys, where earthquakes are common. Scientists have discovered 'black smokers' in the rift valleys. Black smokers are places where hot, mineral-rich water reaches the surface through cracks in the rocks. Minerals are deposited around the cracks to form tall 'chimneys', around which strange creatures live. Molten lava also flows on to the surface in the rift valleys, creating new crustal rock. The other main feature of the oceans are enormous trenches. They are the deepest places in the oceans.

▲ **Around a black smoker**
Hot water bubbles from a chimney on the ocean floor. Minerals feed tiny bacteria, which in turn provide food for giant clams and tubeworms.

Major mountains of the world

Continent	Mountain	Height (in metres)
Africa	Kilimanjaro	5,895
Antarctica	Vinson Massif	5,140
Asia	Mount Everest	8,848
Australia	Mount Kosciusko	2,228
Europe	Mount Elbrus	5,642
North America	Mount McKinley	6,194
South America	Aconcagua	6,959

▲ **A volcanic island**
When an undersea volcano erupts, it may spell the birth of an island, if enough molten rock is flung up.

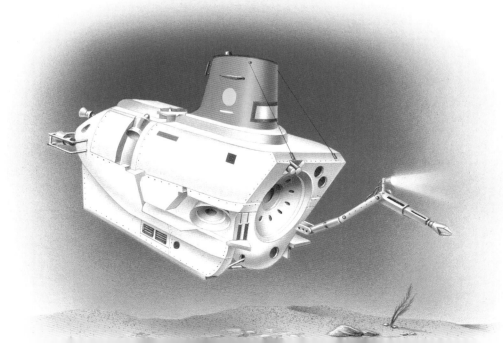

◄ **Intrepid sub**
This submersible craft, called Alvin, has carried scientists down to the dark ocean floor. The scientists took photographs and collected samples. Unmanned subs are also used to explore the ocean bed. They collect video footage for the scientists.

11

The Changing Earth

For millions of years after the Earth was formed, its surface was probably covered by molten rocks. Heavy elements, such as iron, sank to the centre of the Earth to form a solid inner core and a molten outer core. Around the core was a thick mantle, composed of dense (heavy) rocks. Light material rose to the surface, while gases, released from the rocks by volcanic explosions, formed the beginnings of an atmosphere.

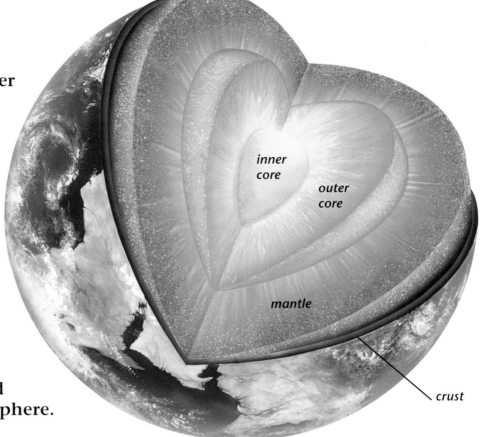

inner core

outer core

mantle

crust

▲ Inside the Earth

The Earth's crust is a thin shell, averaging 35 to 40 km under the continents and 6 km under the oceans. The crust encloses the 2,900-km-thick mantle, the liquid outer core and the solid inner. The core is composed mostly of iron.

Life on Earth

Imagine all of the Earth's history compressed into 24 hours. On this scale, the first-known fossils would have formed at 05.45. However, the first creatures with backbones (the fishes) would not appear until 21.20. The first land plants would have appeared at 21.50 and the first land animals (amphibians) at 22.00. The first dinosaurs would have appeared at 22.50 but they would have become extinct at 23.40. People would have evolved only in the last 40 seconds!

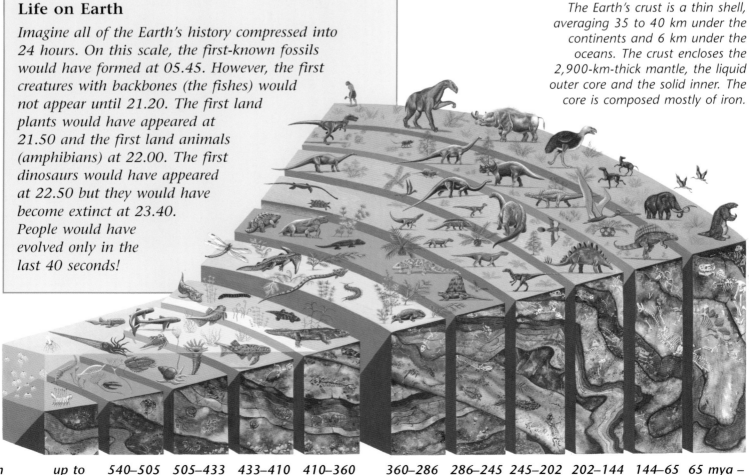

| million years ago (mya) | up to 540 mya | 540–505 mya | 505–433 mya | 433–410 mya | 410–360 mya | 360–286 mya | 286–245 mya | 245–202 mya | 202–144 mya | 144–65 mya | 65 mya – present |

▲ Around 280 million years ago, the world's land formed just one supercontinent, called Pangaea.

▲ By 180 million years ago, Pangaea began to split into Laurasia (to the north) and Gondwanaland (south).

▲ Around 65 million years ago, the Atlantic was widening, India was moving towards Asia, and Australia and Antarctica were still joined.

▲ Today

▲ Continental drift
Over millions of years, plate movements have changed the face of the Earth. These movements are still going on today.

▼ Plate pressure
Earthquakes are most likely to happen where the edges of two plates, that are moving in different directions, grind against each other.

As the Earth cooled, the surface hardened to form a thin crust, while rain from the atmosphere filled hollows to form the first oceans and seas. The oldest rocks found on Earth are about 3.9 billion years old. No one knows when life began, but the oldest fossils of simple organisms found so far were in rocks about 3.5 billion years old.

From a study of fossils, scientists have pieced together the story of life on Earth. They have also discovered that the surface of our planet is always changing. This is because the Earth's hard outer layers, consisting of the crust and the top, rigid layer of the mantle, are split into huge blocks called plates. These plates, which are about 100 kilometres thick, are moved around by currents in the partly-molten mantle beneath them. Plates do not move smoothly. Their jagged edges are locked together for most of the time. But pressure makes the rocks break. They then lurch forward, causing earthquakes.

▲ Earthquake hot spot
The San Andreas Fault, California, is where the Northern Pacific plate and the North American plate are sliding past each other in opposite directions. There have been many serious quakes along this fault, including the devastating ones that hit San Francisco in 1906 and 1989.

The plates do not join up neatly. There are three main kinds of 'join'. Along the rift valleys in the ocean ridges, plates are moving apart. Molten rock, called magma, rises to fill the gaps and new crust is formed.

Along ocean trenches, one plate is sinking beneath another. The edge of the sinking plate, called the subduction zone, melts into magma.

The third kind of plate edge is called a transform fault, such as the San Andreas Fault in the United States. This is a long crack between two plates that are moving alongside each other. Sudden movements along the fault cause severe earthquakes.

Introduction
Shaping the Land

Plate movements create land features. For example, chains of volcanoes form where one plate is sinking beneath another. Also, when two plates carrying land areas collide, the rocks along the plate edges are buckled upwards into folds. Such collisions create long fold mountain ranges, such as the Alps, Himalayas and Rockies. Plate movements also stretch and crack rocks, creating long faults. Sometimes, blocks of rock are pushed upwards between faults to form block mountains. Blocks that sink between faults form steep-sided rift valleys.

The land is also shaped by erosion. For example, when water freezes in cracks in rocks, the ice occupies more space than the water. Constant freezing and thawing gradually enlarges cracks until the rocks split apart. Shattered rocks tumble downhill.

▶ *Many fossil ammonites were formed after the dead animals were buried on the seabed. The remains decayed, leaving holes, or moulds, in the rocks. The moulds were later filled by minerals to form fossil casts.*

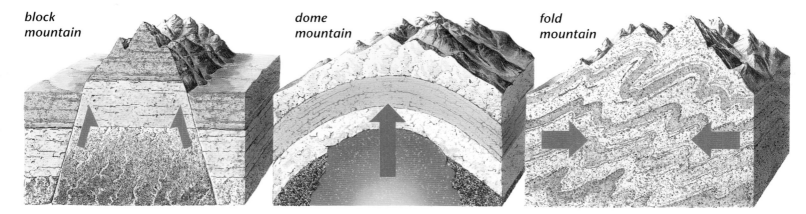

block mountain   dome mountain   fold mountain

▲ **Building mountains**
Plate movements form fold and block mountains. Dome mountains are pushed up by hot molten rock in the Earth's mantle.

▶ **Fold mountains**
The Alps formed when plate movement squeezed rock up into folds.

◀ **The Great Rift Valley**
This valley, stretching from Turkey to Mozambique, was caused by plate movements long ago.

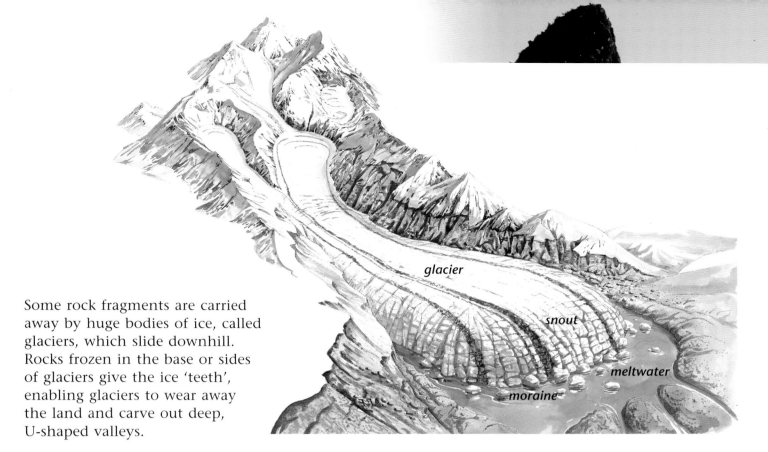

glacier

snout

meltwater

moraine

Some rock fragments are carried away by huge bodies of ice, called glaciers, which slide downhill. Rocks frozen in the base or sides of glaciers give the ice 'teeth', enabling glaciers to wear away the land and carve out deep, U-shaped valleys.

Rivers also carry away rock fragments, breaking them up into small pieces. Rivers wear away V-shaped valleys, because the moving rock fragments on the river-bed rub against the land, loosening and eroding other rocks. Waves and currents also create bays, headlands, cliffs and caves along coasts. In dry areas, wind-blown sand scours rocks, whittling them into strange, fantastical shapes.

Worn fragments of rocks, ranging from pebbles to fine particles of sand and mud, often pile up on the beds of lakes and seas. They squash together to make layers of rocks. These rocks often contain the fossils of ancient creatures which were buried at the bottom of the lake or sea.

Rocks formed from rock fragments are called sedimentary rocks. The other two main kinds of rocks are igneous and metamorphic rocks. Igneous rocks are formed from molten material. Some, such as granite, form when magma cools underground. Others, such as basalt, form when lava hardens on the surface. Metamorphic rocks are rocks changed by great heat or intense pressure. For example, heat and pressure change the sedimentary rock limestone into marble.

▲ **Grand Canyon**
The Yellowstone River has worn a deep, V-shaped valley in the Yellowstone National Park in Wyoming, United States.

▲ **Sedimentary rocks**
Most sedimentary rocks first formed in layers at the bottom of the sea or a lake. Over time, the rocks may be exposed and weathered to form shapely mountains.

▼ **Giant's Causeway**
Basalt, a rock formed on the surface from molten lava, sometimes hardens to form six-sided columns.

▼ **Granite**
The ancient Egyptians mined granite and used it to build statues and other monuments. This igneous rock is hard-wearing and easy to carve.

Weather and Climate

Weather is the state of the air around the Earth. It may be hot or cold, wet or dry, windy or calm. Most of the weather which affects us occurs in the lowest layer of the atmosphere, the troposphere. The troposphere is about 18 kilometres thick above the Equator and about 10 kilometres thick over the poles. It contains more than 75 percent of the air in the atmosphere.

▲ **Weather satellite**
Satellites send back images of cloud systems and other weather information, such as temperatures in the upper air.

Climate

Climate is the usual, or average, weather of a place. The five main climatic types are tropical rainy climates, dry climates, warm temperate (or middle latitude) climates, cold forest climates and polar climates. Each of these is further subdivided according to special features of temperature and rainfall. Some scientists recognize a sixth, mountain climate type, because on mountains temperatures fall the higher you climb. As a result, the tops of high mountains near the Equator are as cold as polar regions.

The atmosphere includes other layers, such as the stratosphere, directly above the troposphere. The stratosphere contains a layer of a gas called ozone, which blocks out most of the Sun's harmful ultraviolet rays.

The atmosphere is always moving. Near the Equator, the Sun heats the ground and hot air rises. The rising air eventually cools and spreads out north and south. Around latitudes 30 degrees north and south, the air sinks back to the surface. Some of this air flows back to the Equator, forming trade winds, while some forms the westerly winds. These are called prevailing (main) winds. The other prevailing winds are the cold polar easterlies that flow from the poles.

▼ **Living communities**
Scientists divide the world into several biomes – plant and animal communities that cover large areas. The largest influence on what lives in a biome is the climate. For example, the treeless tundra is a biome in the north polar region.

▲ **Solar heat**
The Sun's heat is most intense around the Equator. Near the poles, the Sun's rays are spread over a much larger area.

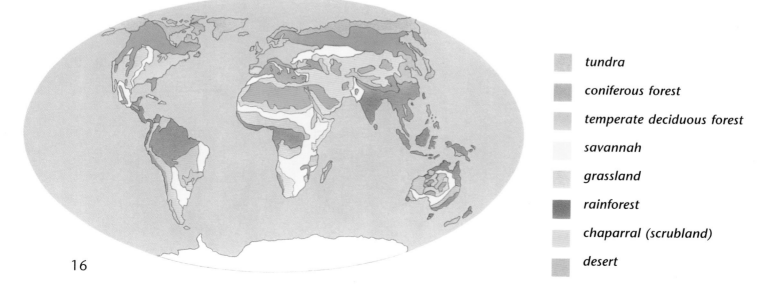

- tundra
- coniferous forest
- temperate deciduous forest
- savannah
- grassland
- rainforest
- chaparral (scrubland)
- desert

The Water Cycle

There is only so much water on the Earth. Because we cannot make new water, clouds and rain are important parts of the water cycle. Heat from the Sun changes water from oceans, lakes, seas, trees, the soil and other moist surfaces into water vapour which forms the rain clouds. Most rain falls directly back into the oceans. The water that falls on land also finds its way back to the oceans in the end, by draining into rivers that lead to the sea. Then it is ready to go round the water cycle all over again.

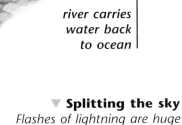

rain cloud

river carries water back to ocean

Air contains moisture in the form of invisible water vapour. Hot air can contain more moist water vapour than cold air. Hence, when warm air rises and cools, it releases the water vapour in the form of droplets of water or ice crystals. Billions of droplets and ice crystals form clouds.

Clouds are the source of rain and snow. Huge cumulonimbus clouds form in thunderstorms. Thunderstorms, which bring heavy rain, are the most common storms. Many occur in the rising air near the Equator. Others occur in the depressions that form along the polar front, the boundary between the cold polar easterlies and the warm westerly winds.

Other storms called hurricanes form over the oceans north and south of the Equator. When hurricanes reach land, they cause tremendous damage. However, the strongest winds of all occur in relatively small storms, called tornadoes.

▶ Prevailing winds
Trade winds, westerlies and polar easterlies are the world's prevailing winds. Around the Equator, winds blow from east to west. Local winds often blow in different directions from the prevailing winds.

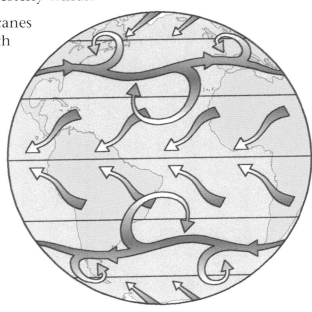

▼ Eye of the storm
The centre of a hurricane, known as the eye, is an area of calm surrounded by violent, swirling winds.

▼ Splitting the sky
Flashes of lightning are huge discharges of electricity in clouds. The intense heat created by lightning causes thunderclaps.

▼ Land and sea breezes
1. During the day, cool winds from the sea blow over the hot land.

2. At night, the land cools more quickly than the water and cool winds blow from the land over the warmer sea.

17

Introduction
Plants
and Animals

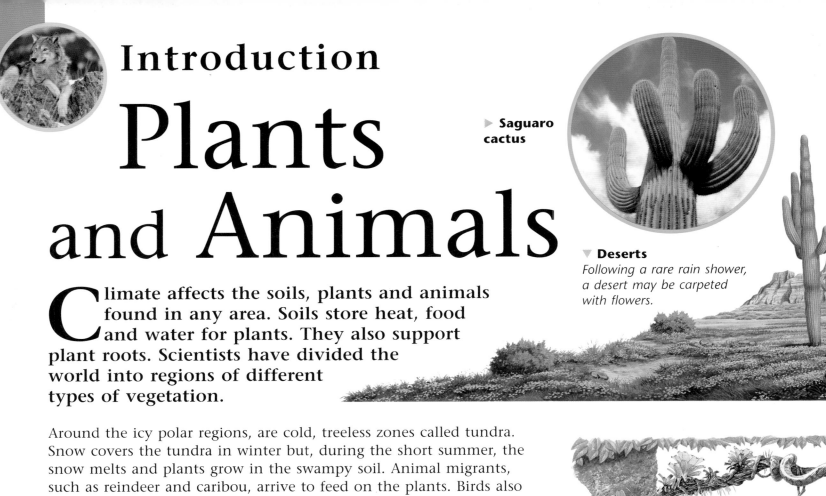

► **Saguaro cactus**

▼ **Deserts**
Following a rare rain shower, a desert may be carpeted with flowers.

Climate affects the soils, plants and animals found in any area. Soils store heat, food and water for plants. They also support plant roots. Scientists have divided the world into regions of different types of vegetation.

Around the icy polar regions, are cold, treeless zones called tundra. Snow covers the tundra in winter but, during the short summer, the snow melts and plants grow in the swampy soil. Animal migrants, such as reindeer and caribou, arrive to feed on the plants. Birds also nest there, feeding on the swarms of insects that fill the air.

There are many different types of forest. In the far north, south of the tundra, contains vast, cold coniferous forests, home to evergreen trees such as fir, pine, larch and spruce. The forests contain such animals as bears, mink, moose and wolves. By contrast, warm temperate regions contain forests of deciduous trees, such as ash, beech, chestnut and oak, that shed their leaves in winter. Much has been cut down to create farmland, though wild boars, deer, foxes and weasels survive in some areas. Mediterranean regions contain large areas of heathland, with their wiry grasses and trees such as cork oaks, myrtles and olives. Tropical rainforests contain more than half of the world's living species. Most wild creatures, including monkeys, snakes, and many birds, live in the leafy canopy.

▲ **Wetlands**
Lakes, rivers and swamps form habitats for many plants and animals. Each species is adapted to its environment.

◄ **Crocodile**

▼ **Grasslands**
Prairies are mid-latitude grasslands which are too dry for forest growth.

▲ **Giraffe, African savanna**

◄ Polar bear

▼ Tundra
During the short summer, the empty tundra becomes a breeding area for many migrating birds. In winter, the ground is frozen.

▲ Angelfish

▼ Coral reefs
The world's coral reefs, which thrive in clean, warm seas, are threatened by human activities, including pollution and tourism.

◄ Tropical rainforests
Hot, steamy rainforests are rich in plant and animal species. Scientists believe that they contain many species yet to be discovered.

▼ Tree frog

Savanna (grassland with scattered trees) borders the rainforests. The African savanna is especially rich in wildlife, including elephants, giraffes, lions, rhinoceroses and zebras. Other kinds of grassland are much drier. They include the prairies of North America and the dry steppes of southeastern Europe and central Asia.

Desert plants and animals must survive long periods without water. For example, cacti in North America's deserts store water in their swollen stems. Similarly, camels can go for several days without drinking.

Mountains have varied plant and animal zones. Only sure-footed animals, such as yaks, ibexes and wild sheep, can scale the steepest, rocky slopes. Other plant and animal regions include lakes and rivers, and the oceans.

► Mountain goat

► Mountains
Temperatures fall at higher altitudes, so plant and animal habitats in mountain regions vary according to the height of the land.

Introduction
Population

▼ **Human evolution**
Human-like creatures evolved on Earth over the last four million years, but modern people appeared only recently.

About 10,000 years ago, life on Earth was hard and most people lived by hunting animals and gathering plant foods. But once people had learned how to farm the land, food supplies increased and so also did the number of people.

By 8000BC, the world population was still only about eight million (today there are cities with more people than that!), but it steadily increased, reaching 300 million by AD1000. The billion mark was first reached in about 1850 and the two billion mark in the 1920s.

In 1975, the world population reached four billion, while the six billion mark was passed in 1999. The rapid increase in the 20th century, called the 'population explosion', is expected to continue.

Homo sapiens sapiens (Modern man)

Homo sapiens neanderthalensis (Neanderthal man)

Homo erectus (Upright man)

Homo habilis (Handy man)

Australopithecus (Southern ape)

| 4–1.5 mya | 2.5–1.5 mya | 1.5 million – 200,000 years ago | 120,000– 30,000 years ago | From 15,000 years ago |

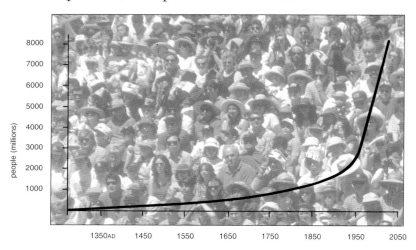

◀ **Population growth**
In 1999, the world's population passed six billion. It is expected to level out at around 11 billion in 2200.

people (millions)

8000
7000
6000
5000
4000
3000
2000
1000

1350AD 1450 1550 1650 1750 1850 1950 2050

▼ **It takes all kinds**
The world's people are an exciting mix of different races, languages and cultures.

Official Languages

From the 14th century, European explorers and, later, colonists spread around the world. In some colonies, the people spoke many languages. When such countries won their independence, they often adopted the language of the former colonial power as their official language when they became independent. As a result, English is now the official language of about 27 percent of the world's population. It is also the language for most international business. Other major official and business languages include Chinese, Hindi, Spanish, Russian and French.

БАНГКОК / BANGKOK
САНКТ ПЕТЕРБУРГ / ST PETERSBURG
ΗΡΆΚΛΕΙΟ / HERAKLION
አዲስ አበባ / ADDIS ABABA
東京 / TOKYO
חיפה / HAIFA
مراكش / MARRAKESH

◀ **Population control**
China has more people than any other country. To control its rapid population growth, the government encourages families to have only one child. Posters explain to people why population control is good for China. But some people oppose this policy. They want many sons to support them when they are old.

Religions of the World

Christianity

Christians believe in one God and in the teachings of Jesus Christ. Christianity is the chief religion of the Americas, Europe and Australia.

Islam

Muslims worship Allah. Islam was founded by the prophet Muhammad in Arabia AD622. It is chiefly practised in northern Africa and parts of Asia.

Hinduism
Hindus worship many gods. Hinduism is one of the world's oldest religions – it began 3,500 years ago. Today, it is the main religion of India.

Buddhism
Buddhism is followed in southern, southeastern and eastern Asia. It developed in India from the teachings of Gautama Buddha (the Enlightened One).

Confucianism and Taoism
Confucianism is based on the teachings of Confucius, a Chinese philosopher born around 550BC. Taoism is another native Chinese religion. It is based on the teachings of Lao Tzu and dates to 300BC.

Judaism
Judaism is an ancient religion which developed in southwest Asia. Israel is the first Jewish homeland to exist since Biblical times.

Sikhism
Sikhism was founded in India in the 1400s by Guru Nanak. It combines elements from Islam and Hinduism. The word 'Sikh' means 'disciple'.

Shintoism

Shintoism is the oldest surviving religion of Japan. Shintoists worship forces in nature, including rocks and trees. About three million people follow Shintoism.

The world's people are divided by several factors, including race, language and religion. All people belong to one species, called Homo sapiens. However, most people identify three main groups: Caucasoids (also called whites); Negroids (or blacks); and Mongoloids, such as Chinese and Japanese. Racial discrimination has caused much conflict between people.

Language and religion also divide peoples. Experts argue about the number of languages in the world today, with estimates ranging between 3,000 and 6,500. The largest language group is the Indo-European family, which includes most European languages, as well as Persian and Hindi in Asia. The second-largest language family is Sino-Tibetan, which includes Chinese.

The world's chief religions are Christianity, with about 1.9 billion followers, Islam, with about 1.1 billion, Hinduism, with 780 million, and Buddhism, which is followed by 324 million people.

▲ **Urban life**
People in cities face many problems, including pollution, traffic jams and high crime rates. Yet cities are growing throughout the world, because city people usually get better jobs and services than their country cousins.

Move to the Cities

In the late 18th century, only about three percent of the world's people lived in urban areas. Cities grew quickly during the Industrial Revolution, as factories provided jobs for many people. By the end of the 20th century, around half of the world's people lived in cities. Experts predict that, by 2015, two-thirds of the world's population will be city-dwellers.

Introduction
How People Live

Communications satellite
Service industries such as television boomed in the 20th century.

U ntil about 200 years ago, most people earned their living by working in agriculture. Farming, which supplies most of the world's food and the raw materials to make clothes and other products, remains the biggest employer of labour. But its relative importance has declined, as manufacturing and service industries have increased in importance. The United States is the world's leading trading nation, followed by Germany, Japan, France and the United Kingdom.

One of the most important industries is mining. It produces fuels, mainly coal, oil, natural gas and uranium for use in nuclear power plants, together with metals and many other materials used in manufacturing industries. When reserves of these run out, we may have to mine other planets or moons.

When resources run out...
Biosphere was an artificial world created in Arizona to study how people might live within carefully controlled environments, such as a colony on the Moon.

Structure of production
Agriculture (red), is still important than industry (yellow) in the economy of countries such as India. But services (blue) provide most income. In the United States, services and industry are far more important than farming.

Saxon farmers
Until about 200 years ago, most people depended on farming. Even today, agricurure employs about half of the world's population.

Industrial revolution
The Industrial Revolution began in Britain in the late 18th century and, during the 19th century, it spread across Europe into Russia, North America and Japan. Today, manufacturing is a major employer. However, in some countries, manufacturing industries are highly efficient and automated. As a result, they now employ fewer people. This has led to an expansion of the number of people in service industries.

India

United States

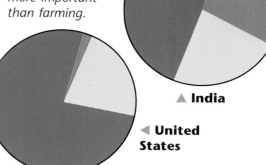

Introduction
Rich and Poor Nations

▲ **Manufacturing**
The world's top manufacturing nations include the United States and Japan.

The modern world contains developed countries, where many people live in luxury, and poor, developing countries, where many people live in slums and often have too little to eat. Experts divide developing countries into those with middle-income economies, and those with low-income economies.

Developed countries with high-income economies import food and raw materials and export manufactured products. By contrast, developing countries with low-income economies have few manufacturing industries and most people work on farms, often at subsistence level. Countries with low-income economies include many African nations, and India and China in Asia. Countries with middle-income economies often export oil or another resource in demand in world trade. Some, such as Argentina, are close to becoming developed countries.

▲ **Subsistence farming**
Many people in Africa and Asia are subsistence farmers. They produce little more than they need to support their families.

▲ **Stock exchange**
The economies of countries have become increasingly involved with each other because of trade and financial links.

Per Capita GNPs

One way of measuring the wealth of a country is to work out its per capita GNP, or gross national product. The per capita GNP is the total value of all the goods and services produced by a country in one year, divided by the population. It is usually expressed in US dollars.

The average per capita GNP in developed countries with high-income economies, such as the United States, is around $23,400. By contrast, the average per capita GNP for developing countries with middle-income economies is $2,500. For developing countries with low-income economies, the average per capita GNP is only $380.

▼ **Gross national product**
The map shows how countries are divided into poor developing economies, middle-income economies and the wealthy high-income economies.

▲ **Mining**
Some under-developed countries have large deposits of oil or valuable minerals. But only a few people benefit and many people remain poor.

Low
$785 or less

Lower middle
$786 - 3115

Upper middle
$3,116 - 9,635

High
$9,636 or more

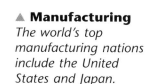

Introduction
The
Environment

◄ **Wind power**
The burning of fossil fuels (coal, gas and oil) causes pollution. Wind turbines can harness the power of the wind to produce 'clean' energy.

The development of the world's natural resources has brought great wealth to many people, especially in developed countries. However, economic development has caused great damage to the natural world. Because of the population explosion, people have flooded into areas of wasteland, cutting down forests and digging up grasslands to create farms. The destruction of plants has led to soil erosion and massive reductions in wildlife. In some areas, once-fertile land has been turned into desert.

The growth of great industrial cities has caused pollution of the land, sea and air. Rivers and lakes have been left lifeless by poisonous industrial wastes; air pollution has affected wildlife and created health problems. Gases released by factories, power plants and cars has caused choking smog in cities. These gases are sometimes dissolved in water droplets in the air. This creates acid rain, which kills trees and most living things in lakes.

◄ **Farewell to the forests**
The destruction of rainforests in South America, Africa and Asia is a major environmental disaster.

▲ **Water pollution**
The release of chemicals into rivers, lakes and seas kills fish and other creatures.

▼ **Filthy air**
Car exhaust fumes pollute the air. Air pollution causes smog which harms people and plants.

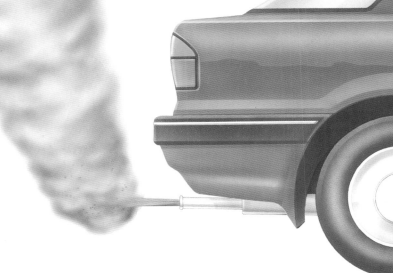

▼ Mountains of rubbish
Unsightly refuse dumps can also be a health hazard. They provide homes for disease-carrying creatures, such as rats.

▶ Solar panels
Energy from the Sun can be harnessed by solar panels. In the future it may replace fossil fuels as a source of environmentally-friendly power.

▼ Recycling
Many waste materials, including metals, glass and paper, can be recycled. Reuse reduces the amount of garbage that must be dumped or burned.

The gases that pollute the atmosphere include carbon dioxide, which is released by the burning of the fossil fuels – coal, oil and natural gas. Carbon dioxide is called a 'greenhouse gas', because it prevents the heat that rises from the ground escaping into space. As the amount of carbon dioxide in the air increases, so also does the temperature of the air. This process is called 'global warming'. Global warming has already begun to change world climates. In the future, it may melt the ice sheets around the poles and raise the level of the sea. Low-lying islands would disappear from the map, while fertile, thickly-populated coastal lowlands will be flooded.

Gases called chlorofluorocarbons, or CFCs for short, have damaged the layer of ozone in the stratosphere, which protects the Earth from the Sun's burning ultraviolet radiation. Since the 1980s, governments have worked together to reduce the release of CFCs into the air to halt this danger. Such international action is required on many fronts to halt the damage being done by human activities to the Earth's fragile environment.

▼ Greenhouse effect
Greenhouse gases in the atmosphere retain heat that is reflected from the Earth's surface. This used to be a helpful thing, but the steady increase of these gases due to human activities has warmed the planet to dangerous levels over the last 150 years.

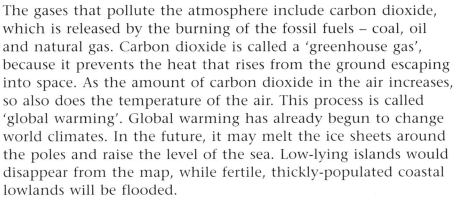

World Facts

THE PLANET EARTH

Age About 4,600 million years
Diameter 12,756 km
Equator 40,075 km
Mass 5,980 billion billion tonnes
Tilt 23.4° from the upright
Distance from Sun 150,000,000 km
Time taken to spin on axis
  23 hrs 56 min
Time taken to orbit Sun
  365 days 6 hrs
Number of moons 1

▲ **Earth seen from Space**

▼ **Tornado, United States**

FACTS

CLIMATE AND WEATHER

Highest-known temperature 58°C
  (Libya, 1922)
Lowest known temperature –89°C
  (Antarctica, 1983)
Wettest place Mawsynram, India
  (average rainfall of 11,873 mm year)
Driest place Atacama Desert, Chile
  (average rainfall 0 mm per year)
Highest-known wind speed
  371km/h at surface level
  (Mt Washington, USA, 1934)
Most powerful tornado 450 km/h
  (Texas,1958)
Heaviest hailstone 1kg
  (Bangladesh, 1986)

FACTS

A RESTLESS PLANET

Worst-known volcanic eruptions
  Thíra, Greece (c.1500BC)
  Taupo, New Zealand (AD130)
  Tambora, Indonesia (1815)
  Krakatoa, Indonesia (1883)
Highest active volcano Ojos del
  Salado, Chile-Argentina (6,887 m)
Most massive active volcano
  Mauna Loa, Hawaii, USA
  (42,500 cu km)
Highest geyser Waimangu,
  New Zealand (460 m, 1903)
Most powerful known earthquakes
  Assam, India (Richter scale 9, 1950)
  Ecuador (Richter scale 8.6, 1906)
  Kamchatka, now Russian Fed.
  (Richter scale 8.3, 1952)
Highest tsunami Alaska, USA
  (524m, 1958)

▲ **Volcano**

EARTH'S LANDSCAPES

Biggest desert *Sahara, North Africa (about 9,269,000 sq km)*

Biggest rainforest *Amazon basin, South America (about 2,500,000 sq km)*

Biggest plateau *Tibetan plateau, Central Asia (about 200,000 sq km)*

Biggest river gorge *Grand Canyon, Arizona, USA (34 9 km long, about 1,615m deep)*

Longest cave system *Mammoth Cave, Kentucky, USA (about 560 km)*

Highest waterfall *Angel Falls, Venezuela (979 m)*

Lowest depression *Lake Vostock, Antarctica (–4,000 m, below ice cap)*

▲ **Desert**

▼ **Nile Delta**

FACTS

MIGHTY RIVERS

1 Nile North Africa (6,695 km)
2 Amazon South America (6,516 km)
3 Chang Jiang China (6,380 km)
4 Mississippi-Missouri USA (6,019 km)
5 Ob-Irtysh Russia (5,570 km)
6 Yenisey-Angara Russia (5,550 km)
7 Huang He China (5,464 km)
8 Congo Central Africa (4,667 km)
9 Paraná South America (4,500 km)
10 Mekong Southeast Asia (4,425 km)

FACTS

OCEANS AND SEAS

1 Pacific Ocean 166,242,000 sq km
2 Atlantic Ocean 86,557,000 sq km
3 Indian Ocean 73,429,000 sq km
4 Arctic Ocean 13,224,000 sq km
5 South China Sea 2,975,000 sq km
6 Caribbean Sea 2,754,000 sq km
7 Mediterranean Sea 2,510,000 sq km
8 Bering Sea 2,261,000 sq km
9 Sea of Okhotsk 1,580,000 sq km
10 Gulf of Mexico 1,544,000 sq km

▲ **Pacific Ocean**

▼ **Mount Everest**

FACTS

SOARING PEAKS

1 Everest or Qomolangma, Himalaya (8,848 m)
2 K2 or Qogir Feng, Himalaya (8,611 m)
3 Kanchenjunga, Himalaya (8,598 m)
4 Lhotse, Himalaya (8,511 m)
5 Yalung Kang, Himalaya (8,502 m)
6 Makalu 1, Himalaya (8,470 m)
7 Dhaulagiri 1, Himalaya (8,172 m)
8 Manaslu 1, Himalaya (8,156 m)
9 Cho Oyu, Himalaya (8,153 m)
10 Nanga Parbat 1, Himalaya (8,126 m)

Countries
of the World

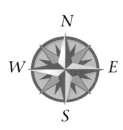

The world is divided into about 190 independent countries, which have control over their own affairs, and more than 40 dependencies, which rely to some extent on an independent country.

The largest independent countries are Russia, Canada, China, the United States and Brazil, while the smallest are Vatican City, Monaco, Nauru, Tuvalu, and San Marino.

The numbers of independent countries and dependencies often change. For example, the former British colony of Hong Kong was reunited with China in 1997.

▲ **An early map**
This map shows the Americas in 1587, during the early days of colonization.

Our Changing World

Here are some of the changes to the world map during the 1990s:

1991 The Soviet Union dissolved and divided into 15 nations

1991–2 The former Yugoslavia split into five countries

1993 Czechoslovakia divided into the Czech and Slovak republics; Eritrea broke away from Ethiopia

1994 Palau in the Pacific became an independent country

1997 Hong Kong reunited with China; Zaire was renamed Democratic Republic of Congo

1999 Portuguese Macao was reunited with China

EUROPE

Europe's western shores face the North Atlantic Ocean. The restless sea has gnawed away at this coastline over the ages, creating chains of islands, headlands and channels. Ocean currents warm the northwest of the continent, giving a moist and mild climate. Southern European lands border the Mediterranean and Black Seas, and are mostly warm and dry. However Europe's northern borders are formed by the ice-bound coasts of the Arctic Ocean. In the east, the continental border with Asia runs overland along the Ural Mountains, the shores of the Caspian Sea and the Caucasus Mountains. Europe and Asia are part of the same vast landmass.

Thousands of years of farming and hundreds of years of industrialization have completely changed the European landscape. Dense woodlands have been cut down and steppes (natural grasslands) have been given over to farming. Great cities and ports have grown up, linked by motorways and railways. Wilderness survives only in a few areas, such as the forests of the far north.

The European Union (EU), currently with 15 member nations, aims to bring increasing economic and political unity to the continent, particularly with the launch of a single currency, the euro. Europe is home to many different peoples, cultures and languages. English, French, and Spanish are now spoken in many other parts of the world.

▲ Bruges, Belgium

▼ Bison

Key to Numbered Countries

1 LUXEMBOURG
2 LIECHTENSTEIN
3 SAN MARINO
4 VATICAN CITY
5 MONACO
6 ANDORRA

ICELAND

SCOT

IRELAND

WALES

PORTUGAL

SPAIN

◄ Grey heron

▼ Oil rig, North Sea

► Mt St Michel, France

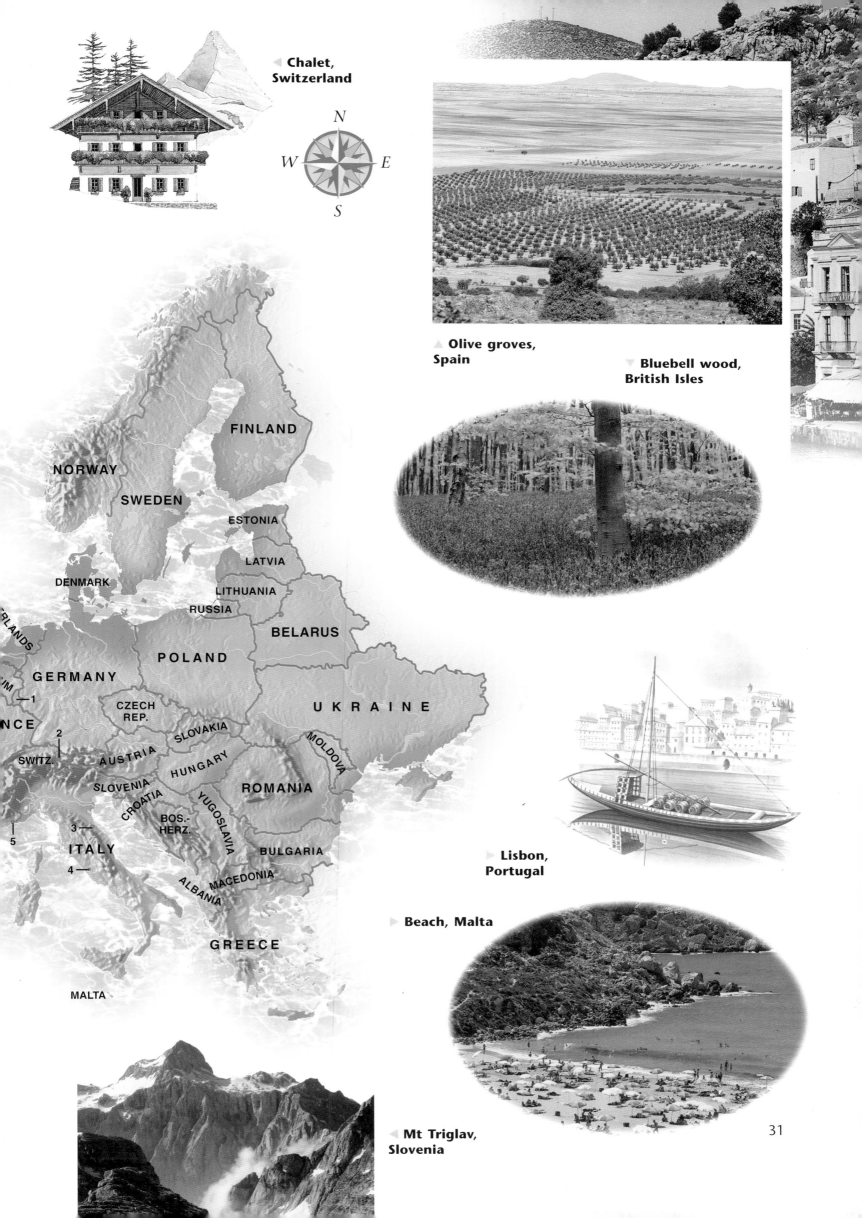

Chalet,
Switzerland

N
W ✦ *E*
S

Olive groves,
Spain

Bluebell wood,
British Isles

FINLAND

NORWAY

SWEDEN

ESTONIA

LATVIA

LITHUANIA

DENMARK

RUSSIA

BELARUS

POLAND

GERMANY

—1

CZECH
REP.

UKRAINE

NCE

2

SLOVAKIA

SWITZ.

AUSTRIA

HUNGARY

MOLDOVA

SLOVENIA

CROATIA

ROMANIA

3

BOS.-
HERZ.

YUGOSLAVIA

5

ITALY

BULGARIA

4 —

MACEDONIA

ALBANIA

GREECE

MALTA

Lisbon,
Portugal

Beach, Malta

31

Mt Triglav,
Slovenia

Scandinavia
and the far north

▲ **Reindeer**

Iceland is a northwestern outpost of Europe, lying just below the Arctic Circle. The island is volcanic, with warm springs and eruptions of lava in a bleak, moorland landscape.

To the east, the Atlantic Ocean extends into the Norwegian and North Seas. The ocean sweeps on through the windy channels of the Skagerrak and Kattegat, into the shallow Baltic Sea. To the south lies Denmark, which takes in the flat lands of the Jutland peninsula and about 500 islands. The biggest of these are Sjælland, Fyn, Lolland and Falster.

To the north lies the Scandinavian peninsula, occupied by Sweden and Norway. This stretches all the way to North Cape, beyond the Arctic Circle, where the sun lights up midsummer nights and the winters remain long and dark. Northern winters are severe but southern regions have a more moderate climate. Lowland plains, covered in birch and spruce, rise to a backbone of rugged mountains. The landscape has been scarred by ancient glaciers. These have created many lakes as well as the deep sea inlets or fiords which indent the Norwegian coast.

Finland lies on the freezing lands between the Scandinavian and Russian Arctic, with coasts on the Gulfs of Bothnia and Finland. Pitted with over 55,000 lakes, this is the most densely forested country in Europe.

▲ **Northern lights**
The aurora borealis (northern lights) flicker eerily across the night sky in Arctic regions. They are caused by particles streaming from the Sun.

▲ **Norwegian fiord**
A fiord is a deep valley that was carved by a glacier during the Ice Age. When the glacier melted, the valley flooded with salty sea water.

▼ **Gushing geysers**
Iceland's geysers are caused by hot rock inside the Earth. The rock heats up pockets of water underground and makes spouts of steam gush up to the surface.

▶ **Wolverine**
The wolverine lives in the forests of the cold north. It is a fierce predator, and will even attack deer or bear cubs.

ICELAND

Grimsey

Raufarhöfn
Kópasker
Isafjördur · Olafsfjördur · Húsavik
Thingeyri · Hólmavik · Saudárkrókur · Vopnafjördur
Vatneyri · Blönduós · Akureyri · Myvatn
Seydisfjördur
Eskifjördur · Neskaupstadur
Olafsvik · Stykkishólmur · ICELAND
Borgarnes · HOFSJÖKULL
Akranes · VATNAJÖKULL
Höfn
Pingvallavatn · Reykjavik
Porisvatn
Keflavik
Stokkseyri · ▲Hekla 1,491 m · ▲Hvannadalshnúkur
MYRDALSJÖKULL · 2,119m
Heimaey
Vestmannaeyjar · Vik
Surtsey

FINLAND

ICELAND

NORWAY

DENMARK

North Cape
Hammerfest · Vadsø
Polmak
Kirkenes
Alta · Utsjoki
Tromsø · Karasjok · Inarijärvi · RUSSIA
Mt. Haltia
▲1,324m
LOFOTEN · VESTERÅLEN · LAPLAND
Narvik · Enontekiö
Kiruna
Mt. Kebnekaise
▲2,111m · Sodankylä
Bodø · Vittangi
Gällivare · Pelkosenniemi
Jokkmokk · Rovaniemi
Tornio · Kemi
Boden
Luleå · Kajaani
Mosjøen · Sorsele · Piteå · Oulu
Storuman · Skellefte · Skellefteå
Ume · Bygdeå
Dorotea · Umeå · FINLAND
Namsos · Grong · Kokkola · Outokumpu
NORWEGIAN · Steinkjer · Ornsköldsvik · Jakobstad · Joensuu
SEA · Ostersund · Kuopio
Trondheim · Vaasa
Kristiansund · Kramfors · Seinäjoki · Jyväskylä
Ålesund · Sundsvall · Tampere
Sunndalsøra · Ljusdal · Port · Kouvola
Dombås · Röros · Hudiksvall · Rauma · Hämeenlinna · Lahti
Galdhøpiggen · Särna · Bollnäs · Söderhamn · Hyvinkää · Kotka
2,469m · Mora · Turku · Helsinki
NORWAY · Lillehammer · Falun · Gävle
Voss · Gjøvik · Borlänge · ÅLAND
Bergen · Västerdal · Mariehamn
Uskedal · Drammen · Västerås · Uppsala
Haugesund · Oslo · Karlstad
Stavanger · Skien · Fredrikstad · Örebro · Eskilstuna · Stockholm
Larvik · Strömstad · Södertälje
Egersund · Arendal · Vänern · Norrköping
Kristiansand · Uddevalla · Vättern · Linköping · GOTLAND
Mandal · Trollhättan · BALTIC SEA
Skagerrak · Göteborg · Jönköping · Västervik · Visby
Boras
Ålborg · Växjö · ÖLAND
Kattegat · Borgholm
Holstebro · Viborg · Halmstad · Kalmar
Randers
JUTLAND · Horsens · Helsingborg · Karlskrona
Esbjerg · Århus · Kristianstad
DENMARK · Malmö
Kolding · Copenhagen · Ystad · Bornholm
Odense · Trelleborg · Rønne
GERMANY

▲ Beaver
Scandinavia is one of the last strongholds of the European beaver. This species was once common in northern lands.

▼ Helsinki port
The Finnish capital has a fine, natural harbour which has to be cleared of ice during the winter months.

▼ Downhill skiing
Both Alpine (mountain) and Nordic (cross-country) skiing are popular during snowy Scandinavian winters. One wooden ski found in Sweden is believed to date back to around 2500BC.

FACTS

DENMARK
Kongeriget Danmark
Area: *43,094 sq km*
Population: *5.3 million*
Capital: *Copenhagen*
Other cities: *Århus, Odense*
Highest point: *Yding Skovhøj (173 m)*
Official language: *Danish*
Currency: *Danish krone*

FINLAND
Suomen Tasavalta – Republiken Finland
Area: *338,145 sq km*
Population: *5.1 million*
Capital: *Helsinki*
Other cities: *Tampere, Turku*
Highest point: *Haltiatunturi (1,328 m)*
Official languages: *Finnish, Swedish*
Currency: *Markka*

Scandinavia and the far north

THE TERM 'SCANDINAVIA' COMMONLY REFERS to three northern European monarchies – Sweden, Norway and Denmark – but it sometimes also includes the republics of Iceland and Finland. All these countries are democracies that have enjoyed a high standard of living in the last 50 years. Denmark, Sweden and Finland are members of the European Union (EU).

Denmark's self-governing overseas territories include the Faeroe Islands and Greenland, which is part of North America. Norwegian territories include the Arctic islands of Jan Mayen and the Svalbard archipelago.

The development of these lands on the rim of Europe has been conditioned by harsh northern winters and by remote and difficult terrain, much of it covered in dense forests. The population is concentrated in the milder southern and coastal regions, and it is here that most farming, industry and trade is located.

▲ Wind turbines
Danish engineers pioneered wind turbine technology and today groups of windmills are a familiar sight in the Danish countryside. They may be up to 60 metres high, generating enough electricity for export as well as for local needs.

▲ Royal guard
A soldier stands guard outside the Amalienborg palace, Copenhagen. The palace is the home of the Danish royal family.

▶ Stockholm
Stockholm, the Swedish capital, lies on a strait between Lake Malar and the Baltic coast. The city's waterfronts also take in a number of small islands linked by bridges.

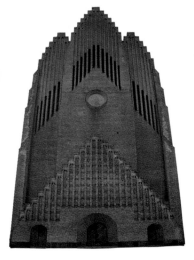

▲ Grundtvig's Church
This modern Danish church is named after Nikolai Grundtvig (1783–1872), who founded the Evangelical Lutheran People's Church of Denmark.

▼ Legoland, Denmark
Denmark exports the most famous toy in the world, Lego. Tourists flock to Legoland, where a tiny model of every major city has been built from the brightly-coloured plastic bricks.

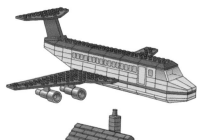

Only eight percent of Finland is suitable for agriculture, and only three percent of Norway. However to the south, Denmark is able to be a major producer of bacon and dairy items such as butter, cheese and yoghurt. In volcanic Iceland, underground heat has been harnessed to warm greenhouses, where flower: and vegetables are grown for export.

The great forests of northern Scandinavia and Finland provide timber for paper-making, construction and furniture. These lands also produce vehicles, machinery, and scientific an electrical equipment. Sweden has rich reserves of iron ore and uranium, and both Denmark and Norway have profited from North Sea oil and natural gas. Northern mountain streams and lakes generate hydroelectric power, and Denmark has been at the forefront of developing wind power. Fisheries have always provided food and a livelihood for each one of these northern lands.

Reykjavík, in Iceland, is the world's most northerly capital city. The capitals of the region are all old seaports which grew up on the trading routes between the Baltic, the North Sea and the North Atlantic.

▲ **Oil from the sea**
Large reserves of oil beneath the North Sea have made Norway Europe's largest oil producer. Norway also has major industries based on natural gas and petrochemicals.

▲ **Timber!**
Timber is an important resource in Norway and Sweden. Some 65 percent of Finland is forested, and wood products make up 40 percent of Finnish exports.

◄ **Salmon farming**
Salmon are farmed in Iceland, where they are an important export. Wild salmon are also caught in the rivers.

FACTS

ICELAND
Lýðveldið Ísland
Area: *103,000 sq km*
Population: *0.3 million*
Capital: *Reykjavík*
Other cities: *Kópavogur, Hafnarfjörður*
Highest point: *Hvannadalshnúkur (2,119 m)*
Official language: *Icelandic*
Currency: *Icelandic króna*

NORWAY
Kongeriket Norge
Area: *323,877 sq km*
Population: *4.4 million*
Capital: *Oslo*
Other cities: *Bergen, Trondheim*
Highest point: *Galdhøppigen (2,469 m)*
Official language: *Norwegian*
Currency: *Norwegian krone*

SWEDEN
Konungariket Sverige
Area: *449,964 sq km*
Population: *8.8 million*
Capital: *Stockholm*
Other cities: *Göteborg, Malmö*
Highest point: *Kebnekaise (2,111 m)*
Official language: *Swedish*
Currency: *Swedish krona*

Scandinavia and the far north

SCANDINAVIA IS HOME TO SEVERAL different peoples. The Saami have lived there since prehistoric times. Their modern homeland, known as Lapland, stretches through Arctic Norway, Sweden and Finland into Russia. Their language is related to Finnish and Estonian. Traditionally, the Saami are nomadic reindeer herders, but many now live in towns and work in other industries.

The Finns make up the majority of people living in Finland. They too speak their own language. Their ancestors moved into the region, from what is now Russia, about 2,000 years ago. About the same time, other peoples were moving north into Scandinavia from Germany. These were the ancestors of today's Swedes (some of whom also live in Finland), Danes, Norwegians and Icelanders. Swedish, Danish, Icelandic and the two spoken versions of Norwegian (Bokmål and Nynorsk) are separate languages which have sprung from similar Germanic roots.

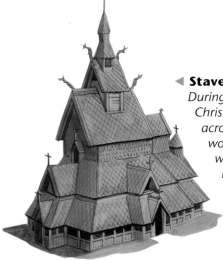

▲ **Saami tribesman**
This Saami from Sweden still wears traditional costume for special occasions. The amount of Arctic pasture available for grazing has been greatly reduced in modern times, and only one in ten Saami now work herding reindeer.

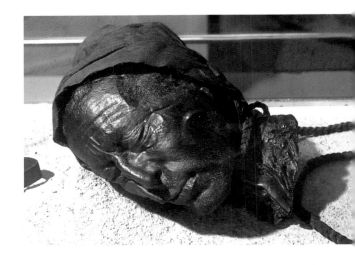

▲ **Tollund man**
Discovered at Tollund Fen, Denmark, in 1950, the Tollund Man is over 2,000 years old. His body has been preserved in a peat bog since Iron Age times. The peat also dyed his skin brown.

◄ **Stave church**
During the 11th century, Christianity began to spread across Norway. About 600 wooden stave churches were built there during the Middle Ages. Today, only about twenty-four remain standing. They are named after their corner-posts, or staves.

▼ **Viking raiders**
The Vikings set sail from their Scandinavian homelands to trade – and also to raid the coasts of Europe. Armed with fearsome iron swords and axes, they looted towns, seized land and took captives.

▲ **A brave leader**
King Gustavus Adolphus of Sweden (1611–32) died after the Battle of Lützen, in the Thirty Years War. The Protestant Swedes were fighting Spain and the Holy Roman Empire.

◄ **Garden of culture**
Tivoli pleasure gardens in Copenhagen provide the venue for open-air theatres, pavilions and a concert hall. In daylight, the flower beds are a sea of colour while at night, coloured floodlights and firework displays brighten

▲ **Viking ship**

About 1,200 years ago the Scandinavians began to take to the sea to seek their fortune abroad. Known as Vikings or Northmen, they attacked coastal settlements all over the British Isles and Western Europe. They sailed westwards to settle Iceland and Greenland, and even discovered North America. They founded new states in Russia and traded with the Arabs in the Middle East.

By the 1100s the Viking homelands had become Christian kingdoms, which grew rich by trading in fish, timber and furs. In 1397 Sweden was united with Norway and Denmark by the Union of Kalmar. Sweden broke away in 1523 and over the next two centuries became a very powerful nation. Norway regained independence in 1905 and Iceland in 1944. Finland was ruled by Russia from 1809–1917. After political division and German and Russian invasions during World War II (1939–45), Europe's far north enjoyed peace and increasing prosperity under a series of liberal governments. Most Scandinavian Christians are Protestants.

AD	**TIMELINE**
c.200	Germanic tribes control most of Scandinavia; Finns settle Finland
c.750	The Viking age (until c.1100): overseas raids, trade, settlement
872	King Harald Finehair of Norway conquers Scandinavia
874	Norwegian Vikings settle Iceland
930	Establishment of Althing, one of the world's oldest parliaments, in Iceland
982	Danes colonize Greenland
1397	Union of Kalmar unites mainland Scandinavia under Danish rule
1523	Sweden leaves the Union
1540	Sweden rules Finland (until 1809)
1618	Thirty Years War (until 1648): Scandinavia supports Protestants
1700	Great Northern War (until 1721): Sweden invades Russia, but is defeated
1809	Russia rules Finland (until 1917)
1905	Norway becomes independent
1919	Finland recognised as independent republic
1920	Iceland becomes independent
1939	World War II (until 1945): Germany invades Denmark and Norway; Sweden remains neutral; Finland invaded by Russia, allies itself to Germany in 1941
1944	Iceland becomes republic
1973	Denmark joins EEC (later EU)
1994	Sweden, Finland vote to join EU; Norway votes against

▲ **Clean design**
Scandinavian design is admired around the world for its unfussy look. This chair by Esko Pajamies was inspired by the renowned Finn, Alvar Aalto (1898–1976).

▲ **St Lucy's Day**
In Sweden, where northern winters are dark and long, it is small wonder that people like to honour St Lucy, a Christian saint associated with light. Her feast day falls on December 13th. Girls dress in white and wear a crown of fairy lights or candles.

37

The Low Countries

The Netherlands, Belgium and Luxembourg lie in northwestern Europe, between France and Germany. They mostly enjoy a mild and moist climate.

The sand dunes of the West Frisian islands and the North Sea coast fringe a low plain. For thousands of years these marshy lands were flooded by the great rivers that spill across them – the Rhine, Schelde and Meuse – and by wind-whipped tides powering in from the sea.

Over the centuries, the peoples of the lowlands learned to build sea and river defences – dykes, dams and barriers. They also became expert at draining marshes and flooded land. The large windmills that powered the drainage pumps may still be seen in many places. Areas reclaimed from the sea, known as polders, now form wide areas of green farmland. A network of canals crosses the countryside.

To the south, the flat lands give way to sandy heaths and fertile plateaus, rising to the Ardennes. This wooded range of hills, some of which rise to over 600 metres, crosses southeastern Belgium, northern Luxembourg and part of eastern France. Southern Luxembourg is taken up by rolling farmland, bordered by the Sûre and Moselle Rivers and crossed by the Alzette.

▲ **Dutch windmill**
As you cycle alongside Dutch canals, you are sure to see beautiful old windmills rising above the fields. Some are still in working order.

▲ **The Amstel River**
Amsterdam is the capital city of the Netherlands. It lies on the Amstel River, which has been channelled into canals to prevent flooding.

▼ **Holding back**
At 32.5km, the Afsluitdijk, between North Holland and Friesland, is the world's longest sea dam. Completed in 1935, it created a huge new lake, the Ijsselmeer.

▲ **The Grand-Place**
At the heart of Brussels, the Belgian capital, is the Grand-Place. Here, flower-sellers show off their wares. Every other year, there is a special floral display, when the square becomes a carpet of begonias.

▲ Space-age landmark
The Atomium was built for the World Fair, held in Brussels in 1958. It was a symbol of scientific progress.

West Frisian Islands
Ameland
Terscheling
Vlieland
Leeuwarden
Groningen
Texel
Waddenzee
Sneek
Assen
Barrier Dam
Emmen
IJsselmeer
Alkmaar
North-East Polder
Meppel
Markerwaard Polder (planned)
Almelo
Zaanstad
Flevoland Polder
Zwolle
NETHERLANDS
Haarlem
Amsterdam
Hilversum
Enschede
Leiden
Amersfoort
Apeldoorn
Ijssel
Gouda
Utrecht
The Hague
Delft
Lek
Arnhem
Rotterdam
Waal
Nijmegen
Dordrecht
Maas
GERMANY
s'Hertogenbosch
Oosterschelde
Breda
Tilburg
Vlissingen
Westerschelde
Eindhoven
Zeebrugge
Venlo
Ostend
Bruges
Antwerp
St. Niklaas
Genk
Heerlen
Roeslare
Ghent
Mechelen
Hasselt
Maastricht
Kortijk
Aalst
Brussels
Leuven (Louvain)
Vaalserberg 321m
Schelde
Waterloo
Liège
Verviers
Tournai
BELGIUM
Huy
Meuse
Spa
Botrange 694m
La Louvière
Namur
Sambre
Mons
Charleroi
Dinant
ARDENNES MOUNTAINS
Buurgplatz 559m
GERMANY
Bastogne
Libramont
LUXEMBOURG
Luxembourg
Esch-sur-Alzette
FRANCE

N
W E
S

NETHERLANDS

▲ Copper butterfly

▼ Forest walks
Built on a bend in the Sûre River, Luxembourg, is the old town of Esch-sur-Sûre. The town attracts many tourists, who can enjoy walks in the spectacular wooded hills that surround the town.

BELGIUM

LUXEMBOURG

▲ Grey heron

FACTS

BELGIUM
Royaume de Belgique – Koninkrijk Belgïe
Area: 30,519 sq km
Population: 10.2 million
Capital: Brussels
Other cities: Antwerp, Ghent
Highest point: Botrange (694 m)
Official languages: Flemish, French
Currency: Belgian franc

LUXEMBOURG
Grand-Duché de Luxembourg
Area: 2,586 sq km
Population: 0.4 million
Capital: Luxembourg City
Other cities: Esch-sur-Alzette
Highest point: Buurgplatz (559 m)
Official languages: French, German, Letzebuergesch
Currency: Luxembourg franc

NETHERLANDS
Koninkrijk der Nederlanden
Area: 40,844 sq km
Population: 15.5 million
Capital: Amsterdam
Other cities: Rotterdam, the Hague
Highest point: Vaalser Berg (321 m)
Official language: Dutch
Currency: Guilder

The Low Countries

BELGIUM, THE NETHERLANDS AND LUXEMBOURG are all small countries, lying in a densely populated part of northern Europe. Together they are sometimes referred to as Benelux, the name of an economic union they formed in 1948. In 1957 the same three countries went on to become founder members of the European Economic Community (EEC), the ancestor of today's European Union (EU). All three are democracies. The Netherlands and Belgium are kingdoms, while Luxembourg is a grand duchy.

▲ **Green and red peppers**

▲ **Trams in Amsterdam**
Amsterdam's trams screech through the old city, crossing rings of canals. Board No 24 from Central Station for Dam Square, the old Mint and the shops of Vijzelstraat.

The two most populous provinces of the Netherlands are North and South Holland, and the whole country is often referred to as Holland. The capital is Amsterdam, an old trading port where tall merchants' houses from the 1600s still line the canals. It is a centre of business and the arts, with a lively youth culture. The government sits at The Hague. The Netherlands is a wealthy country. It produces beers, vegetables and dairy products, being particularly famous for its cheeses. The fields are brightly-coloured with tulips in the spring. Bulbs and cut flowers are important exports. The industrial south produces household and electrical goods. Rotterdam is the world's busiest seaport, sprawling out over 100 square kilometres.

▲ **Tulip bulbs...**
In spring, when the tulips are blooming, many Dutch fields take on the appearance of a child's colouring book. The Dutch passion for tulips dates back to the 1500s.

◄ ▼ **...and light bulbs**
Eindhoven is a major industrial centre in the central southern Netherlands. It is dominated by Phillips, a Dutch-founded multinational firm that specializes in electrical goods for the home and in electronic technology.

▲ **Rotterdam port**
Rotterdam, in the Dutch province of South Holland, was destroyed by bombing during World War II (1939–45). It was rebuilt and today is Europe's most important seaport.

◀ Belgian lacemaker
A Belgian lacemaker, in traditional dress, skilfully moves bobbins across a pillow. Cities such as Bruges have been famous for their lace since medieval times.

▲ Melt in the mouth
Belgium is famous for its luxurious chocolates. Each truffle is handmade, using the finest cocoa.

Belgium has been famous for its cloth since the Middle Ages and is still a major textile producer. It manufactures steel and chemicals, but coal mining is no longer the great industry that it once was. Popular food products include beers, handmade chocolates and cooked meats from the Ardennes. The Belgian capital is Brussels, in the province of Brabant, which also serves as the chief adminstrative centre of the EU.

Tiny Luxembourg is home to both the European Parliament and the European Court of Justice. In terms of income per head, it is the world's richest country. Once a major steel producer, Luxembourg's wealth now comes from banking and finance. Wines are grown along the steep river valleys of the east.

▲ Europe United
Luxembourg City is home to the EU Parliament and Court of Justice. The flags of EU members fly outside.

◀ Belgian beer
Belgium grows more than 400 tonnes of barley every year. Much of it goes to the brewing industry.

▼ Bruges
Known as 'the Venice of the north', the port of Bruges in northern Belgium has many picturesque canals and a thriving tourist industry.

The Low Countries

MOST PEOPLE IN THE NETHERLANDS ARE DUTCH, and the Dutch language is spoken everywhere. The Frisians, who live in the north and on the offshore islands, also speak their own language. Many Dutch cities are home to people whose ancestors came from countries formerly ruled by Holland, such as Surinam and the Dutch East Indies (now Indonesia), as well as workers from other parts of Europe.

The two peoples native to Belgium are the Flemings of the north, who in their language and customs are very close relatives of the Dutch, and the French-speaking Walloons of the south. There have been violent clashes between the two communities over language and culture. French and German are widely spoken in Luxembourg, along with a local language called Letzebuergesch.

In ancient times, this region of northern Europe was occupied by both Germanic and Celtic peoples, and lay on the edge of the Roman empire. In the 700s and 800s, it became part of the Frankish empire. During the Middle Ages, rule passed from one family of European nobles to another. As wars of religion tore Europe apart in the 1500s, the Protestant Dutch fought against rule by Roman Catholic Spain.

▲ **Vincent van Gogh**
The Dutch artist Vincent van Gogh painted this self-portrait after cutting off part of his ear. His landscapes and still-life paintings display the same boldness and original use of colour.

▲ **Language rights demonstration**
Language rights are a live political issue in Belgium. The Flemings of the north speak Vlaams or Flemish, a dialect of Dutch. The Walloons of the south are French-speakers.

▲ **Gouda cheeses**
Round, yellow Dutch cheeses are laid out and weighed at Alkmaar in North Holland – a glimpse of tradition that is always popular with the many tourists that visit the area.

◄ **A trading empire**
The Dutch East India Company was founded in 1602. Its ships sailed to the Dutch East Indies (now Indonesia) to buy coffee, tea and spices. On the way, the merchants would stop off at the Cape of Good Hope in southern Africa, to stock up on supplies for their voyage.

◄ The big race
The Elfstedentocht, or 11 Towns' Tour, dates back to the 1600s. At 200 km, it is the longest ice-skating race in the world. It is also the biggest, with as many as 16,000 competitors. The race is normally held in the Netherlands, but if the weather is too warm, it is held somewhere else.

By the 1600s the Netherlands was a wealthy, independent sea power, trading with the Far East. The year 1806 saw the region under French rule, but independence was soon restored, with Belgium and Luxembourg added to Dutch territory. These two later broke away – Belgium in 1830 and Luxembourg in 1867. The Low Countries suffered great hardship during World War II (1939–45).

Since the Middle Ages, the region has been a centre of scholarship and the arts. Famous philosophers included Erasmus (1466–1536) and Spinoza (1632–77), while great painters included Pieter Breughel (*c.*1520–69), Rembrandt van Rijn (1606–69), Vincent van Gogh (1863–90) and Piet Mondrian (1872–1944).

▶ Football crazy
Dutch soccer teams, such as Ajax, have won fans all over Europe and the national team is always a major contender in World Cup events.

◄ Anne Frank
Anne Frank (1929–45) was a young Jewish girl who spent two years living in a secret annexe in this house in Amsterdam, when the Germans occupied the Netherlands. In 1944, her family was discovered and deported. Anne died in a concentration camp. Her moving diary of the time spent in hiding has been published in 50 different languages.

AD	**TIMELINE**
c.50	*River Rhine becomes border of Roman Empire*
714	*Franks rule most of Low Countries*
922	*Counts of Holland rule Low Countries (until 1384)*
963	*Luxembourg part of Holy Roman Empire*
1354	*Luxembourg becomes a grand duchy*
1384	*Burgundy rules Low Countries, Luxembourg (until 1487)*
1519	*Low Countries part of Habsburg empire*
1568	*Start of Eighty Years War against Spanish rule in Netherlands; Belgium remains Spanish*
1602	*Founding of Dutch East India Company*
1648	*Netherlands recognized as independent*
1700	*France rules Belgium (until 1713)*
1713	*Austria rules Belgium (until 1789)*
1795	*France rules Low Countries (until 1815)*
1815	*United Kingdom of the Netherlands*
1830	*Belgium breaks away from the Netherlands*
1867	*Luxembourg independent under Dutch monarch*
1890	*Luxembourg has its own grand duke*
1914	*World War I (until 1918): Germany invades Belgium; Netherlands neutral*
1939	*World War II (until 1945): Germany invades Belgium and Netherlands*
1948	*Benelux economic union formed*
1957	*Belgium, Netherlands, Luxembourg founder members of EEC (later EU)*
1986	*Completion of Delta Project flood defences*

The British Isles

◀ **European badger**

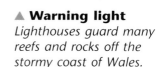

▲ **Warning light**
Lighthouses guard many reefs and rocks off the stormy coast of Wales.

The British Isles rise from a continental shelf, an underwater ledge of land extending into the Atlantic Ocean from northwestern Europe. They include about 5,000 small islands and two large ones, Great Britain and Ireland. Warm ocean currents keep the climate moderate, with winds bringing the heaviest rainfall to western coasts. The British Isles are occupied by two nations, the United Kingdom (a union of England, Scotland, Wales and Northern Ireland) and the Republic of Ireland.

Southern England includes low chalk hills and plains of fertile clay. The river Thames flows into the North Sea, to the south of the flat lands of East Anglia. The rugged peninsula of Cornwall extends southwest towards the Scilly Isles. Northern England includes bleak moors rising to the Pennines, a chain of low mountains and the Lake District of Cumbria.

To the west, Wales is a country of hills, green valleys and mountains, bordering the Irish Sea. Across the Scottish border lie fertile lowlands, the island's highest mountain chains and lakes, known as 'lochs'. Scotland is fringed by the island chains of the Hebrides, Shetlands and Orkneys.

▲ **Bluebell woods**
In early May, large areas of woodland in the British Isles are carpeted with sweetly-scented wild flowers called bluebells.

Ireland is a land of lakes, river estuaries, peat bogs and green farmland, rising to low mountain chains. In the west, tall cliffs and long, sandy beaches meet the Atlantic breakers.

▶ **Lochs and isles**
This castle lies on a tiny island in the middle of Loch Linnhe, on the western coast of Scotland, just south of Ben Nevis.

FACTS

UNITED KINGDOM
United Kingdom of Great Britain and Northern Ireland
Area: *244,100 sq km*
Population: *58.8 million*
Capital: *London*
Highest point: *Ben Nevis (1,343 m)*
Official languages: *English, Welsh*
Currency: *Pound sterling*

REPUBLIC OF IRELAND
Eire
Area: *70,284 sq km*
Population: *3.6 million*
Capital: *Dublin*
Other cities: *Cork, Limerick*
Highest point: *Carrauntoohil (1,041 m)*
Official languages: *Irish, English*
Currency: *Irish pound (punt)*

▲ **The Forth Bridge**
The Forth River flows from the Grampians down to the Firth of Forth. At the estuary, it is crossed by a cantilever railway bridge, built in the 1880s, and a road suspension bridge, completed in 1964.

UNITED
KINGDOM

SCOTLAND

NORTHERN
IRELAND

ATLANTIC
OCEAN

IRELAND

▲ **Red squirrel**

ENGLAND

SHETLAND ISLANDS
Yell · Unst
Foula · Lerwick
Sumburgh Head
Fair Isle

NORTH SEA

Westray
ORKNEY ISLANDS
Kirkwall
Hoy · South Ronaldsay
John o'Groats

Cape Wrath
Butt of Lewis
Thurso
Stornoway
Lewis
North Minch
NORTH-WEST HIGHLANDS
Moray Firth
Fraserburgh
Peterhead
OUTER HEBRIDES
North Uist
Skye
Inverness
Loch Ness
Spey
Don
Dee
Aberdeen
South Uist
Rhum
Mallaig
Barra
Coll
Ben Nevis
1,343 m
GRAMPIAN MTS.
Montrose
Tiree
Mull
Oban
Perth
SIDLAW HILLS
Dundee
Jura
Loch Lomond
OCHIL HILLS
Forth
Firth of Forth
Islay
Greenock
Glasgow
Edinburgh · St. Abbs Head
SCOTLAND
Berwick-upon-Tweed
Holy I.
Kilmarnock
Clyde
Arran
Ayr
Tweed
Jedburgh
NORTH CHANNEL
SOUTHERN UPLANDS
CHEVIOT HILLS
Tory I.
Malin Head
Rathlin I.
Giants Causeway
Newcastle upon Tyne
Aran I.
Londonderry
ANTRIM
Dumfries
Carlisle
Tyne
Sunderland
DONEGAL MTS.
SPERRIN MTS.
Lough Neagh
Stranraer
Solway Firth
Durham
Middlesbrough
Donegal Bay
Donegal
Belfast
Lake District
Scafell Pike
978 m
NORTH YORK MOORS
Scarborough
Erris Head
Lower Lough Erne
Armagh
Isle of Man
Swale
Flamborough Head
Lough Conn
Sligo
Upper Lough Erne
Slieve Donard
852 m
Douglas
Walney I.
Leeds
York
Clew Bay
Lough Allen
Dundalk
Morecambe Bay
Blackpool
Kingston upon Hull
Lough Mask
Lough Ree
Boyne
IRISH SEA
Preston
Bradford
Spurn Head
Lough Corrib
Athlone
Anglesey
Wigan
Manchester
Oldham
Rotherham
Sheffield
LINCOLN WOLDS
Galway
BOG OF ALLEN
Liffey
Dublin
Holyhead
Liverpool
ENGLAND
Trent
Galway Bay
IRELAND
Dun Laoghaire
Llandudno
Wrexham
Derby
Nottingham
The Wash
ARAN ISLANDS
Lough Derg
Carlow
WICKLOW MTS.
Wicklow Head
Caernarfon Bay
Snowdon
1,085 m
Stoke on Trent
Leicester
THE FENS
Norwich
Loop Head
Limerick
Shannon
Nore
Barrow
Bardsey I.
CAMBRIAN MTS.
Wolverhampton
Walsall
Welland
Peterborough
EAST ANGLIA
Gt. Blasket I.
Tipperary
Wexford
Cardigan Bay
Birmingham
Coventry
Northampton
Cambridge
Ipswich
Dingle Bay
Killarney
Blackwater
Waterford
Aberystwyth
WALES
Wye
Cheltenham
COTSWOLD HILLS
Oxford
CHILTERNS
Luton
Colchester
Carrauntoohill
1,041 m
Cork
Hook Head
Cardigan
Milton Keynes
Chelmsford
Kenmare River
Bantry
Old Head of Kinsale
Carmarthen
Severn
Gloucester
Swindon
London
Southend-on-Sea
Bantry Bay
St. Brides Bay
Swansea
Newport
Bristol
Basingstoke
Reading
Thames
Canterbury
Mizen Head
Milford Haven
Gower Peninsula
Cardiff
Bristol Channel
MENDIP HILLS
Salisbury
HAMPSHIRE DOWNS
NORTH DOWNS
Dover
Lundy
Ilfracombe
EXMOOR
Bridgwater
Winchester
THE WEALD
Folkestone
Southampton
SOUTH DOWNS
Hastings
Bude
Exeter
Bournemouth
Portsmouth
Brighton
DARTMOOR
Torbay
Portland Bill
Isle of Wight
ENGLISH CHANNEL
St. Ives
Plymouth
Lyme Bay
Penzance
ISLES OF SCILLY
Lands End
Lizard Point
Alderney
CHANNEL ISLANDS
Guernsey
Jersey

WALES

▼ **A road into the sea**
The rocky pillars of the Giant's Causeway, on the coast of Northern Ireland, are formed from a volcanic rock called basalt.

▶ **Green and pleasant land**
The county of Yorkshire in northern England has a landscape of rolling dales (valleys) and bleak moors.

The British Isles

THE UNITED KINGDOM (UK) BRINGS TOGETHER THE TWO KINGDOMS OF England and Scotland, the principality of Wales and the province of Northern Ireland. All are ruled by a single monarch and have a long history of democracy. In 1997 both Wales and Scotland voted to have their own parliaments. The UK is a member of the European Union (EU). The Isle of Man and the Channel Islands are self-ruling states associated with the UK.

The most densely populated area is southeastern England, linked by rail tunnel with northern France. The capital, London, is home to about 6.8 million people. In the 1800s many large industrial cities also grew up in the English Midlands and in the North. The traditional industries of these regions, such as coal, textiles, and shipbuilding have all declined. The English economy today is based upon finance, services, chemicals and telecommunications.

Wales, with its capital at Cardiff, is sheep-farming country. The south is industrialized, but now depends on foreign-owned factories producing goods such as electronic components rather than its coalfields.

▲ Crown jewels
The Crown Jewels of the British kings and queens include a dazzling array of gold and precious stones dating back over 1,000 years.

▲ Big Ben

▼ Welsh farming
Wales has three times as many sheep as people. However, farmers find it increasingly difficult to make a living in the bleak uplands.

▲ Liverpool skyline
Liverpool is a lively city on the river Mersey, in northwestern England. During the 1800s and early 1900s it became one of the busiest seaports in the world.

▼ Blackpool lights
Sitting on the northwest coast of Lancashire is England's most popular seaside resort, Blackpool. Its attractions include the Blackpool Tower, the famous illuminations (lights) and many amusement parks.

▼ All aboard!
Southern England and northern France, separated by the waters of the Channel about 10,000 years ago, were finally linked by a rail tunnel which opened in 1994.

Scottish trawlers
At Mallaig harbour, fishing boats return to unload the day's catch. Scottish trawlers bring in over 65 percent of all the fish and shellfish caught in the United Kingdom.

Salmon leap
Scotland is also famous for its freshwater fish. Wild salmon head upriver each year to their spawning grounds.

The Scottish capital is the city of Edinburgh, but heavy industry has always been based around Glasgow and the River Clyde. Today's most important resource is oil, produced offshore in the North Sea. Important exports include salmon and Scotch whisky.

Northern Ireland has the industrial city of Belfast as its capital. The province has seen long years of conflict between those who wish to remain within the UK and those who wish to be part of the Republic of Ireland. A peace agreement was reached in 1998.

The capital of the Republic of Ireland is Dublin, on the river Liffey. Ireland's green fields are ideal for dairy farming, while the coasts and rivers provide fishing. Irish industries include brewing, glass-making and computer manufacture. With a long history of poverty, the Irish economy has been transformed in recent years by membership of the EU.

Prime beef
The Aberdeen Angus is a breed of cattle from northeastern Scotland, prized for its tasty meat. Its long coat protects it from the bitter cold.

FACTS

ENGLAND
Area: 130,360 sq km
Capital: London
Other cities: Birmingham, Manchester, Liverpool

NORTHERN IRELAND
Area: 14,150 sq km
Capital: Belfast
Other cities: Derry, Armagh

SCOTLAND
Area: 78,750 sq km
Capital: Edinburgh
Other cities: Glasgow, Aberdeen

WALES
Area: 20,760 sq km
Capital: Cardiff
Other cities: Swansea, Wrexham, Bangor

SELF-GOVERNING ISLANDS
Isle of Man, Guernsey (with six dependencies), Jersey

Milking time
A cow is milked by hand on a small Irish farm. Irish dairy products include delicious butter and cheeses.

Peat bog
Over many thousands of years, large peat bogs have formed in Ireland. This waterlogged turf is cut and dried for fuel in the home and for use in power stations.

The British Isles

THE ENGLISH ARE DESCENDED FROM THE many different peoples who settled in the southern part of the United Kingdom over the ages – British Celts, Germanic peoples such as the Angles, Saxons and Danes, and French-speaking Normans. The English language developed from Anglo-Saxon, but includes many words based on Latin and Norman French. It is the language of some of the world's greatest works of literature, such as the plays of William Shakespeare (1564–1616).

The Welsh and Cornish are mostly the descendants of British Celts and are closely related to the Bretons. They have their own languages. Welsh is spoken by about half a million people in Wales and has a literature dating back about 1,400 years. Festivals of music and poetry called *eisteddfodau* are held throughout Wales, where there is a strong tradition of choral singing.

▲ **Cowes Week**
Cowes, on the Isle of Wight off southern England, is the setting for an annual yacht race, the Fastnet Cup.

▲ **Welsh arts festivals**
At an eisteddfod, poets, musicians, dancers and actors compete for honours.

► **The game of cricket**
The game of cricket was invented in England in the 1700s. It spread from there to Australia, New Zealand, India and Pakistan, South Africa and the Caribbean.

▼ **Globe Theatre**
The Globe was originally raised in 1599, on London's South Bank. Many of Shakespeare's greatest plays were first performed there. Now a new Globe has been built.

▼ **Into the next century**
Under construction at Greenwich, London, the Millennium Dome is a building for the 21st century. Its teflon-coated skin is fixed to 12 steel masts, each 100 metres tall.

▲ Towering angel
Unveiled by artist Antony Gormley in 1998, the Angel of the North rises 20 metres into the sky. It is the largest sculpture in Britain.

▼ Chinatown
Chinese people see in the Chinese New Year in style. Cities such as London, Liverpool and Manchester have their own Chinese districts.

▶ National flowers
Each country in the British Isles has its own national flower. There is the English rose (1), the Scottish thistle (2), the Welsh daffodil (3) and the Irish shamrock (4).

Scots are descended from both Gaelic and British Celts, from a people called the Picts and from Norse invaders. The Scottish version of the Gaelic language may still be heard in the Highlands and islands, while the Anglo-Scots dialect of the Lowlands is regarded by many as language in its own right. It was the language of the great poet Robert Burns (1759–96).

The Manx people are of mixed Celtic and Norse descent, while the Channel Islanders have English, Norman and French roots.

The Irish are mostly descended from Gaelic Celts, along with some Viking, Norman, English and Scottish ancestry. The Irish language, a version of Gaelic, is still spoken in some rural areas. Irish authors and poets have used both the Irish and English languages to great effect.

Many other peoples have settled the British Isles over the ages, contributing to the richness and variety of its culture. They include Roma (gypsies), Jews, Italians, Greek and Turkish Cypriots, Afro-Caribbeans, Chinese, Indians, Pakistanis and Bangladeshis.

▲ Manx cat
The Manx is a tailless breed of cat from the Isle of Man.

▼ A Hindu temple
Britain has many Hindu and Sikh temples, Islamic mosques and Jewish synagogues in addition to its Christian churches.

▲ Goddess of the Celts
The Irish have a strong tradition of story-telling. Many myths date back to the days of the Celts. This statue shows the Celtic horse goddess, Epona, who was worshipped in Europe around 400BC.

The British Isles

ABOUT 2,000 YEARS AGO MOST OF GREAT BRITAIN occupied by a group of Celtic tribes, known as Britons. Picts lived in parts of Scotland, while Ireland was the home of another Celtic group, the Gaels. The Romans invaded in 55 and 54BC and again from AD43. They failed to conquer the far north and Ireland, but stayed for nearly 400 years.

After the Romans left, the Britons came under attack from the west by groups of Gaels known as Scots, and from the south and east by Angles and Saxons – Germanic peoples who founded the small kingdoms which later became England. They gradually drove the Britons into Wales, Cornwall and across the Channel to Brittany. Wales, Scotland and Ireland flourished as centres of Celtic Christian civilization.

Large areas of Britain and Ireland were soon invaded by Vikings and then by Normans. The kingdom founded in England in 1066 became very powerful, ruling large parts of France. It waged constant war with the Welsh and Scots.

▲ **Warrior queen**
Boudicca was queen of a Celtic tribe called the Iceni. In AD60 she led a devastating uprising against the Romans, capturing London and killing 70,000 of the enemy.

▲ **Sacred stones**
Massive stones, lined up with the midsummer sunrise, were raised at Stonehenge in southern England between 3200 and 2000BC.

◀ **Domesday Book**
In 1086 King William I (c.1028–87) ordered a great survey of people and property in the lands he had conquered 20 years earlier. It was used for legal and taxation purposes until 1522.

▲ **The Bayeux Tapestry**
In 1066 William, Duke of Normandy, defeated King Harold of England near Hastings. The story of the Norman invasion was recorded in the Bayeux Tapestry.

◀ **The 1500s**
Henry VIII (1491–1547) ruled England and Wales. He founded the Protestant Church of England. He was a jovial but very ruthless king. He married six times.

▼ **Raleigh's England**
During the reign of Queen Elizabeth I, Britain became a world power. Sir Walter Raleigh (1552–1618) was, for a time, the queen's favourite courtier. He founded the American colony of Virginia. He fell from favour and was beheaded in 1618.

◀ **Death of a king**
In 1649 the whole of Europe was shocked when Parliament ordered the execution of King Charles I, after the Civil War. Britain returned to being a monarchy in 1660, when Charles II took the throne.

▲ **Eamon de Valera**
Eamon de Valera (1882–1975) took part in the Easter Rising in Dublin in 1916. His political party, Sinn Fein, wanted self-government for Ireland.

Wales was formally annexed by England in the 1500s, and the Scottish and English thrones were united in 1603. The new United Kingdom was Protestant. It led savage invasions and settlement of Roman Catholic Ireland, which joined the union in 1800. By then the United Kingdom was becoming the world's first industrialized nation. It ruled a vast overseas empire, from India to Africa and Australia, which gave it raw materials and new markets.

Not until the following century did the power of the British empire begin to fade, exhausted by two world wars (1914–18 and 1939–45). Southern Ireland became a Free State in 1922 and a fully independent republic in 1949.

During the 1960s most of Britain's remaining colonies became independent. By the 1970s the UK and the Irish Republic were looking forward to a more peaceful future, based on co-operation with the other countries of Europe.

▲ **Power to the north**
After nearly 30 years of political strife in Northern Ireland, a peace agreement was reached in 1998. Power was devolved to the region through a new Assembly.

◀ **Edinburgh castle**
Edinburgh, the Scottish capital, was settled over 5,000 years ago. In 1997, the Scots voted to have their own parliament based in Edinburgh.

AD	TIMELINE
43	Roman occupation of Britain (until 446)
795	Start of Viking raids on British Isles
844	Kenneth MacAlpin unites Picts and Scots
1066	Norman invasion of England
1175	England claims Ireland
1284	England conquers Wales
1338	Hundred Years War (until 1453) between England and France
1536	(and 1542) Acts of Union, England-Wales
1608	Scotland and England unite
1649	Republican commonwealth until 1660
1707	Act of Union, England-Scotland
1776	American colonies declare independence
1801	Act of Union, Great Britain-Ireland
1914	World War I (until 1918): against Central powers (including Germany, Austria)
1917	Easter Rising, Ireland
1922	Irish Free State (republic 1949)
1939	World War II (until 1945): against Axis powers (including Germany, Japan)
1973	UK and Ireland join EEC (later EU)
1997	Scots and Welsh vote for own devolved parliaments
1998	Peace agreement, Northern Ireland

France

▶ **Wild boar**

▲ **A snowy peak**
Mont Blanc in the Alps is western Europe's tallest mountain. A 12-km long road tunnel was dug beneath the mountain in the 1960s.

France lies at the centre of western Europe, between Germany and Spain. To the north it is bordered by the sandy shores and chalk cliffs of the Channel coast and the rocky headlands of Brittany. In the southwest, pine forests and dunes face the stormy Atlantic Ocean. The sun-baked hills of southern France descend to the Mediterranean Sea.

Along the Spanish border, the snowy peaks of the Pyrenees form a high barrier. The Alps, which straddle the border with Italy and Switzerland, soar to 4,807 metres at Mont Blanc. Northwards lie the forested slopes of the Jura, Vosges and Ardennes.

Most of northern France is a rolling plain, given over to farming. It is crossed by the winding river Seine. The volcanic rocks of the Massif Central dominate south-central France, to the west of the Rhône valley. The river Rhône flows south to form a delta on the southern coast, creating the wetland region of the Camargue. The Mediterranean island of Corsica is also French territory. It is mountainous, with olive groves and dry scrubland, called maquis.

▲ **Water carrier**
Built 2,000 years ago by the Romans to carry water to Nîmes, the Pont du Gard aqueduct is in southern France.

Monaco is a tiny state on the Mediterranean coast, surrounded by France. It is a built-up zone, which includes new land reclaimed from the sea.

▼ **Island of the monks**
Mont-Saint-Michel lies just off the coast of Normandy. Crowning the island is a Benedictine abbey, that was built in AD966. Below it are medieval houses and fortified walls.

▲ **Insect-eater**
The hoopoe, with its striking crest, is found in southern Europe. Its curved bill is tailor-made for teasing insects out of the bark of trees.

FACTS

FRANCE
République Française
Area: *551,500 sq km*
Population: *58.8 million*
Capital: *Paris*
Other cities: *Marseille, Lyon*
Highest point: *Mont Blanc (4,807 m)*
Official language: *French*
Currency: *French franc*

MONACO
Principauté de Monaco
Area: *1.0 sq km*
Population: *0.03 million*
Capital: *Monaco*
Other cities: *Monte Carlo*
Official language: *French*
Currency: *French franc*

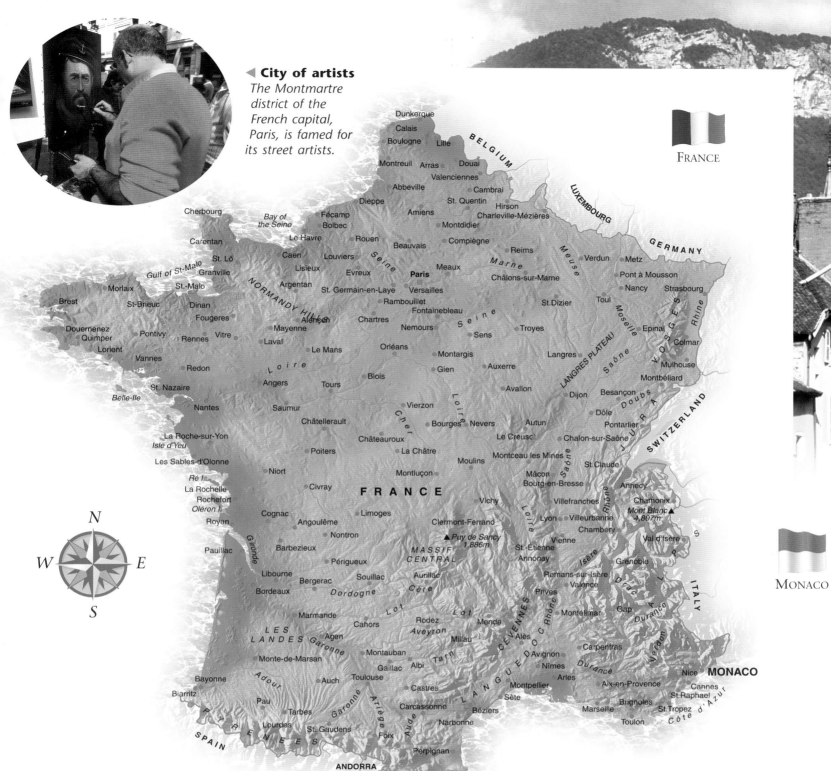

◀ City of artists
The Montmartre district of the French capital, Paris, is famed for its street artists.

FRANCE

MONACO

Dunkerque
Calais
Boulogne
Lille
BELGIUM
Montreuil
Arras
Douai
Valenciennes
Abbeville
Cambrai
LUXEMBOURG
Dieppe
St. Quentin
Hirson
Charleville-Mézières
GERMANY
Amiens
Montdidier
Cherbourg
Bay of the Seine
Fécamp
Bolbec
Rouen
Beauvais
Compiègne
Reims
Verdun
Metz
Carentan
Le Havre
Meaux
Châlons-sur-Marne
Pont à Mousson
St. Lô
Caen
Louviers
Marne
Nancy
Strasbourg
Gulf of St-Malo
Granville
Lisieux
Evreux
Paris
Versailles
St. Dizier
Toul
Morlaix
St.-Malo
Argentan
St. Germain-en-Laye
Meuse
Moselle
Brest
St-Brieuc
Dinan
Fougeres
Rambouillet
Fontainebleau
Seine
Epinal
Colmar
Douernenez
Quimper
Pontivy
Rennes
Vitre
Mayenne
Alençon
Chartres
Nemours
Sens
Troyes
Langres
LANGRES PLATEAU
Mulhouse
Lorient
Vannes
Laval
Le Mans
Orléans
Montargis
Auxerre
Dijon
Besançon
Montbéliard
Redon
Loire
Angers
Tours
Blois
Gien
Avallon
Saône
Doubs
VOSGES
Rhine
St. Nazaire
Saumur
Vierzon
Loire
Bourges
Nevers
Autun
Dôle
Pontarlier
JURA
SWITZERLAND
Belle-Ile
Nantes
Châtellerault
Cher
Châteauroux
La Châtre
Le Creusot
Chalon-sur-Saône
St. Claude
La Roche-sur-Yon
Isle d'Yeu
Poitiers
Moulins
Montceau les Mines
Mâcon
Annecy
Les Sables-d'Olonne
Niort
Montluçon
F R A N C E
Bourg-en-Bresse
Chamonix
Mont Blanc ▲ 4,807m
Ré I.
La Rochelle
Rochefort
Civray
Vichy
Villefranches
Villeurbanne
Oléron I.
Royan
Cognac
Limoges
Clermont-Ferrand
Lyon
Chambery
Val d'Isère
Angoulême
Nontron
▲ Puy de Sancy 1,886m
St-Etienne
Vienne
Rhône
Pauillac
Barbezieux
MASSIF CENTRAL
Annonay
Grenoble
ALPS
Gironde
Périgueux
Aurillac
Romans-sur-Isère
Isère
Libourne
Bergerac
Souillac
Valence
Drac
Gap
Bordeaux
Dordogne
Cère
Prives
Montélimar
Durance
ITALY
Marmande
Lot
Rodez
Mende
Lot
Durance
Monte-de-Marsan
Agen
Garonne
Cahors
Aveyron
Millau
Alès
Avignon
Carpentras
Verdon
Nice
MONACO
LES LANDES
Montauban
Gaillac
Albi
Tarn
CÉVENNES
Nîmes
Aix-en-Provence
Cannes
St. Raphael
Bayonne
Auch
Toulouse
Castres
LANGUEDOC
Arles
Marseille
Brignoles
St. Tropez
Côte d'Azur
Biarritz
Pau
Tarbes
Garonne
Carcassonne
Béziers
Montpellier
Sète
Toulon
Lourdes
St. Gaudens
Ariège
Foix
Narbonne
Adour
PYRENEES
Aude
Perpignan
SPAIN
ANDORRA

▼ Loire River
Straddling the Loire is the castle, or château, of Chenonceaux. The Loire is France's longest river. It stretches for 1,020 kilometres from the Cévennes Mountains to the coast of Brittany.

Cape Corse
Bastia
Gulf of Sagone
CORSICA
Ajaccio
Bonifacio
Strait of Bonifacio

▲ Esterel, southern France
The southern coast of France borders the Mediterranean. The inviting blue waters and sunny climates make the coast a popular tourist destination.

▶ The lap of luxury
One of the world's smallest countries, Monaco, attracts very wealthy visitors. The port of its capital, Monaco, is crammed with luxury yachts from around the world.

France

FRANCE IS A DEMOCRATIC REPUBLIC AND A FOUNDER member of the European Economic Community (EEC), now the European Union (EU). It is one of the world's wealthier countries, although it has suffered from economic problems and strikes have been common. Through its former overseas colonies, France still has considerable international influence.

The capital, Paris, has a population of over nine million people. It is located in the centre of northern France, on the river Seine. France's second city is Marseille, a large seaport on the Mediterranean. The country is divided into 22 regions.

The coalfields of northern France have mostly run out. To generate power for industry, France depends heavily on nuclear power and also on hydroelectric schemes. Alternative energy sources include the sun and tides. French factories produce chemicals, aircraft, cars and high-speed locomotives.

Traditionally, France is an agricultural country, and many people still work as farmers. Normandy produces apples and dairy products, and the wide range of climates makes it possible to grow many kinds of crops, from wheat and maize to rice.

▲ High-speed train
Short for train à grande vitesse, the TGV is a feat of French engineering. This high-speed train can top 500 kilometres per hour.

▲ Apples

▲ Fashion capital
A model at the Chanel catwalk show in Paris. Chanel was founded by the great French couturier, Coco Chanel (1883–1971).

▼ Soft cheeses

▶ Tidal barrier, Rance
France is a major exporter of electricity. Most is generated in nuclear power stations, but the French have led the way in trying to find greener ways to make electricity. Here at St Malo, a massive tidal barrier harnesses the power of the river Rance.

▼ Into the future
At Futuroscope, a theme park and high-tech college near Poitiers, French architects were given a free reign to create bold, imaginative designs to inspire the next millenium.

◀ Sunflowers
In the fields of southern France, bright yellow sunflowers ripen in the sun. The seeds of the flowers are harvested and then pressed to make sunflower oil.

▶ Fattening up
Pâté de foie gras is a rich pâté made from goose liver. The birds are fattened up on plenty of corn. Every region of France has its own pâté recipe.

▼ Champagne

The warm hillsides of Provence provide flowers for the perfume industry, while the Loire, Burgundy and Champagne regions are renowned for their fine wines. Forests in many areas provide timber and there are large fishing fleets in the west.

French cooking is generally rated to include some of the best dishes in the world – indeed, French is the language of cooking. The catwalks of Paris are the ultimate achievement of any career as a fashion model. Tourists bring wealth to France too, visiting the Alpine ski resorts, the castles or châteaux of the Loire and the beaches and yachting marinas of the south.

Monaco uses French currency. Although it is such a small country, its tax laws attract many wealthy residents. It is a centre of finance, tourism – and gambling, in the famous casino at Monte Carlo.

▼ Fine wines
The region of Bordeaux produces some of the world's finest red wines. Many vineyards are attached to a stately home, or château, which gives the wine its name.

▶ Fit for a princess
Monaco is ruled by a prince. Prince Rainier, the current monarch, married the glamorous Hollywood actress Grace Kelly.

▼ Fast cars
Motor racing is an important source of income to Monaco. Each year, two important races are held there: the Monaco Grand Prix and the Monte Carlo rally.

55

France

THE FRENCH ARE DESCENDED FROM MANY different roots, including the ancient Celtic people known as Gauls, and the Franks, a Germanic people. The French language mostly comes from Latin, the language of the Roman empire. It is spoken everywhere in France, although there are great historical differences between French as spoken in the north and the Provençal version of the south.

Since the Middle Ages, the French language has been celebrated by poets, playwrights, storytellers and philosophers. France has also produced great philosophers, painters, film-makers and musical composers. Playwrights such as Corneille (1606–84), Molière (1622–73) and Racine (1639–99), painters such as Monet (1840–1926) and Dégas (1834–1917), and composers such as Debussy (1862–1918) are admired around the world.

Within the French borders there are other peoples, besides the French. The Basques and Catalans of the southwest have their own languages and traditions. The homelands of both these peoples stretches across the border into Spain. The Corsicans have their own language and along the eastern borders are speakers of Italian and German.

▲ **Enlightened times**
During the 1700s, France was a centre of learning. The writer Denis Diderot (1713–84) compiled a massive 28-volume encyclopedia, which aimed to further all knowledge.

▲ **A great writer**
The French poet Victor Hugo (1802–85) is best remembered for his novels – Nôtre-Dame de Paris ('The Hunchback of Notre-Dame') and Les Misérables.

▲ **A treasure-trove of art**
The Louvre museum, Paris, houses one of the world's greatest art collections. In the 1980s, a stunning pyramid of steel and glass was built at the entrance to the old museum.

▼ **Military training**
France has a military tradition dating back to the Middle Ages. Most young men must spend some time in the army as military service.

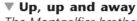

▼ **Up, up and away**
The Montgolfier brothers, Joseph and Jacques, pioneered the use of hot-air balloons in the 1780s.

▲ **Father of comedy**
Jean Baptiste Poquelin (1622–73) was known as Molière. He wrote many plays that were performed for King Louis XIV.

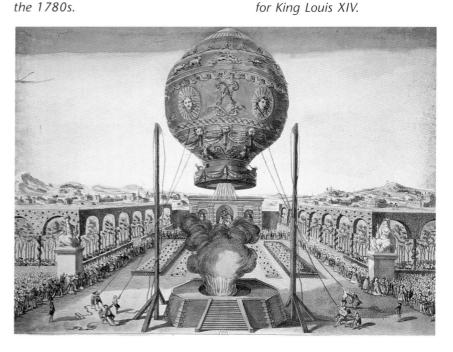

SPEAK... FRENCH

good day	bonjour
goodbye	au revoir
thank you	merci
please	s'il vous plaît
yes	oui
no	non
sorry	pardon
happy birthday!	bon anniversaire!
school	école
children	enfants
milky coffee	café au lait

▶ **Sacré Coeur**
Overlooking the French capital, is the domed Roman Catholic church of Sacré Coeur ('sacred heart'). Although France does not have an official state religion, most French people follow the Catholic faith.

The Bretons of the northwest are descended from British Celts who settled the region about 1,500 years ago. Their language is related to Welsh and Cornish. The Normans are descended from Viking invaders who were granted the region of France that became known as Normandy.

Also living in France there are Roma (gypsies) and Jews. North Africans, descended from the citizens of France's former colonies, make up three percent of the French population. Many of these are Muslims, but nine out of ten French people are Roman Catholics.

A hundred years ago the French were mostly a nation of country-dwellers. Today nearly three-quarters of them live in towns and cities, leaving many rural areas deserted or given over to holiday homes and tourism.

▲ **The naughty nineties**
In the 1890s Paris had the reputation of being a city of fun, of wild dances such as the can-can, of pretty girls, artists and poets.

▲ **A game of bowls**
Pétanque is a popular game played all over France; most towns have a square of gravel set aside for the game. Players aim to bowl their ball as close as possible to a smaller ball.

▼ **Tour de France**
Every year, France hosts the world's most famous cycle race, the Tour de France. The gruelling course stretches over 4,000 kilometres and takes about three weeks to complete.

France

FRANCE HAS BEEN INHABITED SINCE PREHISTORIC times, and cave paintings discovered at Lascaux date back to 15000BC. The ancient Greeks founded the port of Marseille in about 600BC, when most of the country was home to a group of Celtic tribes, known as the Gauls.

The Gauls were conquered by the Romans in 58–48BC, and the Latin language became widely spoken. Germanic-speaking peoples from beyond the river Rhine began to attack the Roman empire. By AD476 it had collapsed and power passed to the Franks, after whom the modern country is named.

The Franks beat back invasion attempts by the Moors (Muslim Berbers and Arabs) and founded a great Western European empire under Charlemagne (768–814). Vikings took control of Normandy in 911, but the French kings became very powerful during the later Middle Ages. They battled with their neighbours, especially the English, and expanded the frontiers of their kingdom. From the Middle Ages until the death in 1715 of Louis XIV, the 'Sun King', France was seen as the centre of European civilization, despite a series of fierce religious wars.

▲ Ice Age art
There are about 600 cave paintings at Lascaux. They show the animals that prehistoric people hunted – bison, stags and horses.

▲ Standing stones
Sited at Carnac, on the coast of Brittany, over 3,000 stone monuments are arranged in avenues and circles. They were erected in Stone Age times.

▲ Into battle
Fierce Frankish armies defeated the Romans, Gauls and Visigoths in their quest for land. By AD540, they controlled most of Roman Gaul.

▲ Burnt at the stake
After hearing voices from God, Joan of Arc led the French into battle in the Hundred Years War. The English burnt her as a witch in 1431.

▶ Medieval France
A code of fine manners, known as chivalry, grew up in the castles and great houses of France in the Middle Ages.

◀ **Louis XIV (1638–1715)**
The 'Sun King' ruled France for 72 years, at the height of French power in Europe. He built the glorious palace of Versailles and spent vast sums on wars.

▲ **Off with their heads!**
By the 1780s, the French had tired of paying exhorbitant taxes to their kings. In the Reign of Terror that followed the Revolution, over 17,000 aristocrats were put to death, including King Louis XVI and his queen.

France began to build up a large empire overseas, but injustice at home led to the outbreak of the French Revolution in 1789. Out of the ruins of this dream of freedom arose a Corsican general called Napoleon Bonaparte, who was crowned emperor in 1804. He led brilliant military campaigns from Spain to Russia and Egypt, but was finally defeated by British and Prussian forces at Waterloo, in 1815.

◀ **The emperor!**
From his revolutionary beginnings, Napoleon Bonaparte (1769–1821) became the most powerful ruler in Europe. He was a great soldier and law-maker.

France then suffered a series of invasions by Germany – in 1870, in 1914 and in 1940. After World War II, France lost most of its overseas empire, but joined its neighbours and former enemies in encouraging economic and political union in Western Europe.

▶ **General de Gaulle**
Charles de Gaulle (1890–1970) led the French resistance to the Nazi occupation in World War II. He later became president of France.

◀ **We will remember them**
To this day, the fields of northeastern France are planted with poppies in remembrance of the soldiers who died there in the trenches during World War I. Millions of Allied and German soldiers lost their lives.

AD	TIMELINE
560	Frankish kingdom covers most of France
911	Vikings granted duchy of Normandy
1297	Grimaldi dynasty rule Monaco
1337	Hundred Years War against England (until 1453)
1491	France and Brittany united
1643	Birth of Louis XIV, France becomes major European power
1789	Start of the French Revolution
1804	Napoleon Bonaparte crowned emperor
1815	Napoleon defeated by British and Prussians at Waterloo
1848	France becomes a republic again
1852	France returns to being an empire
1870	Prussia defeats France Revolution of the Paris Commune The Third Republic
1914	World War I against Germany (until 1918)
1939	World War II against Germany (until 1945): France occupied 1940; Liberation 1944
1957	France founder member of EEC (later EU)
1958	Charles de Gaulle becomes president
1968	Uprisings by students and workers
1969	De Gaulle resigns

Germany and the Alps

The Alps form the highest mountain chain in western Europe. Switzerland occupies the western end of this range. It is a land of lakes, forests, high meadows and icy summits. To the east, across the river Rhine, lies the tiny state of Liechtenstein. The great Alpine peaks of Austria descend to fertile plains around the river Danube.

The Alps also extend northwards into Bavaria, in southern Germany. The Black Forest covers the southwest of the country, and the Bohemian Forest the southeast. The busy river Rhine winds northwards from the French border, to be joined by the Main at Mainz and the Moselle at Koblenz. It flows through steep valleys, whose slopes are covered in vineyards.

From the central Harz Mountains, the German landscape drops to sandy heath and flat farmland, part of the great plain which stretches eastwards into Poland and Russia. Germany's eastern border lies on the rivers Oder and Neisse. Northern Germany, crossed by the rivers Weser and Elbe, lies on the windy coasts of the North Sea and the Baltic. These are linked by the Kiel Canal, which cuts through Schleswig-Holstein on the Danish border.

The climate of the region varies greatly, being generally mild in the west, with harsh winters in the Alps and across the northeastern plains.

▲ Faithful friend
The German Shepherd dog is brave and loyal, which makes it an ideal breed for police work. Originally from Germany, it is now found worldwide.

▲ Austrian Tirol
The province of Tirol in western Austria is a region of lush mountain pastures. In winter, these are buried under a thick blanket of snow.

▼ Alpine pinnacle
The Matterhorn (4,478 metres) rises on the border of Switzerland and Italy. From the Swiss side it looks more like a peak than a ridge.

◀ Mountain goat
The goat is hardy enough to survive on the cold, snowy slopes of the Alps. It produces rich milk and strong-tasting meat.

▲ A royal residence
Overlooking the capital city, Vaduz, this castle is home to the prince of Liechtenstein. Parts of the castle date back to the 1500s.

◀ Nature's palette
Named after its dark spruces, the Black Forest also includes deciduous trees, that put on a colourful show each autumn.

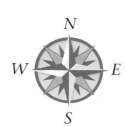

N
W E
S

FACTS

AUSTRIA
Republik Österreich
Area: *83,859 sq km*
Population: *8.1 million*
Capital: *Vienna*
Other cities: *Graz, Linz, Salzburg*
Highest point: *Gross Glockner (3,797 m)*
Official language: *German*
Currency: *Schilling*

GERMANY
Bundesrepublik Deutschland
Area: *356,980 sq km*
Population: *81.9 million*
Capital: *Berlin*
Other cities: *Hamburg, Munich, Cologne*
Highest point: *Zugspitze (2,963 m)*
Official language: *German*
Currency: *Deutsche Mark*

LIECHTENSTEIN
Fürstentum Liechtenstein
Area: *160 sq km*
Population: *0.03 million*
Capital: *Vaduz*
Highest point: *Vorder-Grauspitz (2,599 m)*
Official language: *German*
Currency: *Swiss franc*

▲ Edelweiss

▼ Lake Geneva
Known in French as Lac Léman, this large lake is fed by the river Rhône. It is bordered by France and western Switzerland.

◀ Along the Rhine
The source of the river Rhine is high in the Swiss Alps. Flowing northwest, it passes through Germany and the Netherlands. It is Europe's busiest river. Here, the wine-producing town of Bacharach overlooks the river.

61

Germany and the Alps

SWITZERLAND IS MADE UP OF small regions called cantons. Traditionally, it is a neutral country, which means that it does not take part in any wars or join any international military or political alliances. However it is the headquarters of many international organizations, such as the Red Cross and the World Health Organisation. These are based in its westernmost city, Geneva. Switzerland is a wealthy country. Its income comes from international banking, tourism and the making of instruments and tools as well as chemicals and dairy products such as cheese.

Liechtenstein is a democratic principality, which has close links with Switzerland. It shares the same currency and is also a centre of banking and winter sports. Austria is a democratic republic and has been a member of the European Union (EU) since 1995. It too has a successful tourist industry, and manufactures textiles and chemicals. The mountain slopes provide timber for the mills and water for hydrolectric power schemes. Many tourists visit the capital, Vienna, one of Western Europe's great historical cities.

▲ Mountain homes
Timber-built houses, called chalets, may be seen all over the Alps. They have broad roofs to protect against winter snowfall.

▲ Alpine herds
In spring, when the snows melt, cows are led up to graze the high Alpine pastures. In autumn they return to farms in the valleys.

▲ Rows of beet
Sugar beet is a major crop in this part of Europe. Traditionally, farms in southern Germany and the Alps are small family businesses, while on the plains of eastern Germany they are often very large.

▲ Healing waters
Austrians relax in a cloud of steam. Spas (towns built around hot springs) are hugely popular as health resorts in Germany and Austria.

▲ Swiss gold
Bars of gold are stacked up in the vaults of a Zurich bank.

◄ Birth of the Web
At the European Laboratory for Particle Physics (CERN), in Switzerland, a scientist tests magnets for a new accelerator. One spin-off of the research here was the development of the Worldwide Web.

FACTS

SWITZERLAND
Eidgenossenschaft Schweiz – Confédération Suisse – Confederazione Svizzera

Area: 41,284 sq km
Population: 7.1 million
Capital: Bern
Other cities: Zurich, Geneva
Highest point: Monte Rosa (4,634 m)
Official languages: German, French, Italian
Currency: Swiss franc

◀ **Swiss watches**
Switzerland has long been famous for its manufacture of clocks, watches, precision instruments and machine tools. Traditional cottage industries adapted rapidly to new technology and innovation.

▲ **Compact discs**
Germany today is a major exporter of CDs, high-quality audio systems, radios and electrical goods for the home, such as refrigerators.

Germany is a democratic republic, with a federal system of government. This means that its regions, called *Länder*, have considerable local power. Until 1871 Germany was made up of many different small countries and it was divided again from 1949 until 1990.

Germany has the strongest economy in Europe and was a founder member of the European Economic Community (today's European Union or EU). It produces electrical goods, optical instruments, chemicals and cars, and is world-famous for company names such as Bayer, Bosch, Volkswagen, Mercedes and BMW. German regions are known for their beers, white wines and sausages.

Germany has many large cities, such as Hamburg, Munich, Cologne and Frankfurt am Main, which is a centre of banking and international trade. With Germany re-united, Berlin has once again become the capital of the whole country.

▲ **Eau de Cologne**

▲ **Fruit of the vine**
Vineyards were planted in Germany in Roman times. The chief wine-growing areas are in the southwest and the sunny valleys of the Rhine and Moselle rivers.

▼ **Anti-nuclear protest**
German demonstrators protest against the transportation of nuclear waste. Environmental issues came to the fore in German politics between the 1970s and 90s, with a party called the Greens gaining wide support.

▼ **The Volkswagen**
In German, Volkswagen means 'people's car'. Founded in 1936 to produce cheap, tough family cars, the company had worldwide success from the 1960s onwards. Its works at Wolfsburg remain the world's largest.

Germany and the Alps

ABOUT 65 PERCENT OF THE SWISS PEOPLE SPEAK a regional dialect of German, while 18 percent speak French. In the high mountain valleys of the south you may hear Italian or a language called Romansch. This is a conservative country, which is proud of its independence and traditions. Geneva was one of the birthplaces of the Protestant faith, but Switzerland today also has many Roman Catholics.

Little Liechtenstein is largely Roman Catholic and German-speaking, with most people living in the capital city, Vaduz. Austria too is German-speaking and its countryside is dotted with fine old castles, Catholic churches and monasteries. The mountainous countryside of the Tirol region, in the west, is famous for crafts such as woodcarving, as well as for its folkdances and yodelling. Austria's cities have a long tradition of lively art, theatre and music – Wolfgang Amadeus Mozart (1756–91) and Johann Strauss the Younger (1825–99) were both Austrian.

▲ Sound of the Alps
The alpenhorn, a long-necked, wooden horn, was originally played by cowherds. Its booming sound echoed across mountain valleys and down to the village below.

▲ Folk costume
These dancers wear the traditional costume of Bavaria, in southern Germany. The man wears lederhosen (leather shorts) and the woman a dress called a dirndl.

SPEAK... GERMAN

hello	guten Tag
goodbye	auf Wiedersehen
thank you	danke
please	bitte
yes	ja
no	nein
school	Schule
children	Kinder

▶ Catholic Austria
About 300 years ago, the devout Roman Catholic faith of Austria inspired architects to erect beautiful churches and monasteries. The ornate building-style is known as baroque.

▼ The young Mozart
The great Austrian composer Mozart was born in Salzburg in 1756. He made his first European tour at the age of six. He studied in Italy and eventually became court musician to Emperor Joseph II in Vienna.

▲ On the piste
Skiing in the Alps was once the hobby of a few dedicated enthusiasts. Today, the winter resorts are a major industry, attracting tourists from all over Europe.

Oberammergau
In 1633 a terrible plague struck this village in Bavaria. The survivors vowed to thank God by staging a play about the sufferings of Christ, with all the parts played by the villagers. It is still put on every 10 years.

Clever horses
The days of the Habsburg emperors are recalled at Vienna's Spanish Riding School, where white Lippizaner horses perform elegant exercises to music.

The standard form of the German language, called *Hochdeutsch*, is spoken throughout Germany, but there are also many regional dialects which vary greatly from north to south. In the last forty years or so, many workers from southern Europe and Turkey have also settled in German cities. The north is a stronghold of the Protestant faith, while the south is largely Roman Catholic. There are many traditional festivals and customs, with carnival being celebrated as a major festival from Cologne southwards into Bavaria and Austria.

Germany has a long cultural history, having been the birthplace of the painter Albrecht Dürer (1471–1528), of the composer Beethoven (1770–1827) and of great writers such as Goethe (1749–1832) and Schiller (1759–1805). In the 1800s Germany was also the home of many philosophers and of political thinkers such as Karl Marx (1818–83).

Life observed
Fine detail characterizes the drawings of Albrecht Dürer, the greatest German artist of his day. Born in 1471, Dürer produced hundreds of paintings, engravings and woodcuts.

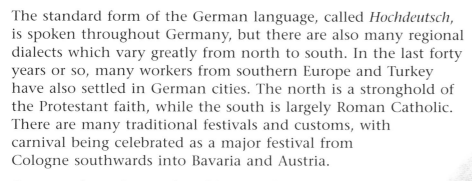

Sigmund Freud
This Austrian psychologist (1856–1939) developed a new method of investigating the unconscious and treating problems of the mind. It was called psychoanalysis.

St Nicholas
December 5th is the Christian festival of St Nicholas. In the Lucerne region of Switzerland, this day is marked by a lantern procession and the exchange of gifts. St Nicholas is the patron saint of children.

Karl Marx
Born at Trier in 1818, Marx showed how history is driven by economic forces and called upon workers to seize power. His ideas inspired communists long after his death in 1883.

Germany and the Alps

THIS REGION WAS ONCE THE HOME OF Celtic tribes, but by about 550BC they had mostly been driven out by Germanic peoples from the north and east. The Romans only succeeded in conquering parts of Germany, founding cities such as Cologne. It was Germanic tribes who finally broke the power of Rome in AD476.

▲ On the river Salzach
The Austrian city of Salzburg has a fascinating history. It has a medieval castle, and a cathedral built 1614–88.

Germany lay within the Frankish empire ruled by Charlemagne (742–814). Although this soon broke up, it was followed by a new Holy Roman Empire which lasted from 962 to 1806. This was a patchwork of small kingdoms, principalities and cities. They fell under the overall rule of the Austrian Habsburg family. Switzerland broke away from Habsburg rule between 1291 and 1499, but Austria did gain control of a vast area of Central Europe.

▲ Celtic art
This Celtic figurine with a wheel symbol was found at Hochdorf in Austria. It dates back to the 500s BC.

In the 1500s, parts of Germany became Protestant, following the teachings of a monk called Martin Luther. Bitter religious wars followed. The Thirty Years War (1618–48) devastated most of the country. During the 1700s power shifted northwards, as a state called Prussia built up a large, modern army. By 1871 the king of Prussia, Wilhelm I, was emperor of a united Germany.

▼ Printing pioneer
Johannes Gutenberg was a pioneer of printing with movable type. He was born at Mainz, Germany in 1400. He moved to Strasbourg and set up a printing press there, perhaps as early as 1439.

▶ The Protestant
Martin Luther (1483–1546) was a German monk who condemned corruption in the Roman Catholic Church and wrote a new translation of the Bible. He led the Protestant Reformation.

▲ A patriotic tale
Legend has it that in 1307, Albert II of Austria tried to gain control of Switzerland. William Tell, an archer from Uri, refused to pledge his loyalty. As a punishment, he was forced to shoot an apple from his son's head. No harm was done and Tell went on to free his country.

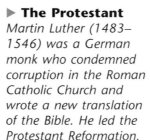

▶ The Thirty Years War
In 1618 a violent quarrel broke out between the Protestants in Bohemia and their Austrian rulers, who were Roman Catholic. This rapidly grew into a religious war which engulfed most of western Europe. It was one of the most savage wars Europe had ever known. Starvation and atrocities were common. Peace came with the Treaty of Westphalia which recognized the rights of Protestants in Germany.

◀ **Maria Theresa**
Born in 1717, Maria Theresa became empress and one of the wisest rulers of Austria. She died in 1780.

▶ **Otto von Bismarck**
Bismarck was a powerful Prussian statesman who engineered the Franco-Prussian War and created a united Germany in 1871.

▼ **Dresden bombed**
As World War II drew to a close, German cities were relentlessly bombed by Allied aircraft. The city of Dresden, with its fine old buildings and priceless works of art, was destroyed in a great firestorm. Up to 130,000 people were killed in the air-raids.

During World War I (1914–18) Germany and Austria suffered disastrous defeat. Germany became a republic, but communists and national socialists ('Nazis') fought each other on the streets. By the 1930s the Nazis were in power. They annexed Austria and German troops invaded much of Europe, during World War II (1939–45). Their leader, a racist dictator called Adolf Hitler, ordered the murder of millions of Jews. Nazi Germany was defeated by the Allies and lay in ruins.

It was next divided into two countries, a Federal Republic in the west, supported by the United States and Western Europe, and a communist Democratic Republic in the east, supported by the Russians. Germany was not reunited until 1990.

▼ **End of the wall**
Germany was divided after World War II. East Germans built a wall across Berlin in 1961 and shot anyone trying to escape. It was finally knocked down in 1989.

▲ **Heil Hitler!**
Adolf Hitler was born in Austria in 1889. Embittered by World War I, he followed violent and racist right-wing politics. He became leader of the Nazi Party and dictator of Germany, murdering millions.

AD	TIMELINE
1282	*Rise of Habsburg dynasty in Austria*
1291	*Swiss struggle against Habsburg rule (until 1499)*
1517	*Luther starts the Protestant Reformation*
1618	*Thirty Years War (until 1648).*
1740	*War of the Austrian Succession (until 1748)*
1805	*France defeats Austria at Austerlitz*
1806	*France defeats Prussia at Jena*
1834	*Prussia unites with small German states*
1848	*Revolution in Germany; failed 1849*
1870	*Prussia defeats France*
1871	*Wilhelm I of Prussia, emperor of united Germany*
1914	*World War I (until 1918): Central powers (Germany, Austria-Hungary) defeated; Switzerland neutral*
1933	*Hitler becomes German Chancellor*
1938	*Germany annexes Austria*
1939	*World War II (until 1945): Axis powers (Germany, Italy and allies) invade much of Europe; defeated by Allies; Switzerland neutral.*
1949	*Germany divides into Federal Republic (West, capitalist) and Democratic Republic (East, communist)*
1957	*German Federal Republic founder member of EEC (later EU)*
1961	*East and West Berlin divided by fortified wall (until 1989)*
1990	*Germany reunified*

Iberian peninsula

▲ **White storks**

▲ **In the Pyrenees**
The Pyrenees Mountains run from the Bay of Biscay to the Mediterranean coast, a distance of 440 kilometres. They include rushing streams and high peaks which reach 3,404 metres above sea level.

The Iberian peninsula is a great block of mainland Europe which juts out into the Atlantic Ocean. It is narrowly separated from North Africa by the Strait of Gibraltar, and its southern shores border the Mediterranean Sea.

The peninsula is largely mountainous. The snow-capped Pyrenees run along the border with France. The Cantabrian ranges, which include the Picos de Europa, run parallel with the north coast, while southern ranges include the Sierra Nevada. Much of the central region is a sun-baked plateau called the Meseta. Major European rivers rise in the mountains. They include the Ebro, Guadalquivir, Duero, Tagus and Guadiana.

▲ **Pardel Lynx**

▲ **Barbary ape**
A legend says that when the apes leave Gibraltar, so will the British.

Most of the Iberian peninsula is taken up by Spain. Portugal occupies the southwestern part, facing the Atlantic Ocean. Andorra is a little state high in the Pyrenees, while Gibraltar, a sheer rock guarding the entrance to the Mediterranean, is a British colony.

Spain also takes in the Balearic island chain in the Mediterranean, the volcanic Canary Islands (geographically, a part of Africa), and two ports on the coast of Morocco. Portugal governs Madeira and the Azores.

Although the northern Iberian coast is green and moist, most of the central region is hot and dry, with parts turning into desert.

▲ **Terraced farmland**
Irrigation and soil conservation are essential in the hot, dry regions of Portugal.

◄ **The Rock**
You can see the African coast from the 426-metre high Rock of Gibraltar.

FACTS

ANDORRA
Principat d'Andorra
Area: 453 sq km
Population: 0.1 million
Capital: Andorra la Vella
Highest point: Coma Pedrosa (2,946 m)
Official language: Catalan
Currency: French franc, Spanish peseta

PORTUGAL
República Portuguesa
Area: 91,982 sq km
Population: 9.9 million
Capital: Lisbon
Other cities: Porto, Setúbal
Highest point: Estrela (2,963 m)
Official language: Portuguese
Currency: Escudo

SPAIN
Reino de España
Area: 505,992 sq km
Population: 39.3 million
Capital: Madrid
Highest point: Pico de Teide (Canary Islands, 3,718 m)
Official language: Spanish
Currency: Peseta

DEPENDENCIES
Gibraltar (UK)

▲ **Burros**

▶ **Split town**
The old town of Ronda in Spain's Andalucia region is divided by a river gorge, El Tajo. Sheer rock faces drop for 130 metres. The two halves are linked by a stunning bridge.

PORTUGAL

ANDORRA

Bay of Biscay

Cape Ortegal
La Coruña
El Ferrol
Carballo
Villalba
Oviedo
Gijón
Llanes
Santander
Cape Finisterre
Lugo
Fonsagrada
CANTABRIAN MOUNTAINS
Bilbao
San Sebastian
Santiago de Compostela
Sarria
Reinosa
Vitoria
Pamplona
PYRENEES
FRANCE
ANDORRA
Pico de Aneto 3,404m
Andorra la Vella
Lalin
León
Astorga
Osorno
Burgos
Logroño
Arga
Figueras
Vigo
Orense
SIERRA CABRERA
Villada
Soria
Saragossa
Gerona
Costa Brava
Baltar
La Gudina
Palencia
Ebro
Gállego
Cinca
Llobregat
Manresa
Braga
Bragança
Valladolid
Zamora
Duero
S P A I N
Jalón
Caspe
Lérida
Tarrasa
Barcelona
Mogadouro
Vila Real
Medina del Campo
Jalón
Reus
Tarragona
Costa Dorada
Porto
Lamego
Douro
Tormes
Segovia
SIERRA DE GUADARRAMA
Tajuña
Tajo
Morella
Vinaroz
Tortosa
Cape Tortosa
PORTUGAL
Salamanca
Avila
Guadalajara
Alcalá de Henares
Teruel
Mijares
Aviero
Viseu
Guarda
Cuidad Rodrigo
Bejar
SIERRA DE GREDOS
Madrid
Aranjuez
Cuenca
Castellón de la Plana
Costa del Azahar
Menorca
Mahón
Coimbra
Covilhã
Plasencia
Tajo
Toledo
Turia
Sagunto
Mallorca
Manacor
Castelo Branco
Leiria
Tomar
MONTES DE TOLEDO
Requena
Valencia
Júcar
Gulf of Valencia
Palma
BALEARIC ISLANDS
Caldas da Rainha
Tagus
Cáceres
Trujillo
Villarrobledo
Alcira
Cape Neo
Ibiza
Santarém
Portalegre
Daimiel
Manzanares
Albacete
Almansa
Alcoy
Ibiza
Formentera
Lisbon
Badajoz
Don Benito
Ciudad Real
Guadiana
Valdepeñas
Alcaraz
Yecla
Alcoy
Alicante
Costa Blanca
Setúbal
Évora
Almendralejo
Puertollano
Elche
Pozoblanco
SIERRA MORENA
Orihuela
Ardila
Azuaga
La Carolina
Linares
SIERRA DE SEGURA
Moratalla
Cehegín
Murcia
Costa Blanca
Beja
Guadiana
Constantina
Córdoba
Jaén
Martos
Segura
Lorca
Cartagena
Cape Palos
Nerva
Guadalquivir
Baza
Aguilas
Chança
Huelva
Seville
Puente Genil
Osuna
Genil
Guadix
Granada
Huércal Overa
Lagos
Faro
Las Marismas
Morón de la Frontera
Antequera
Mulhacén 3,478m
Almería
Cape Gata
Costa Blanca
Cape Saint Vincent
Algarve
Gulf of Cadiz
Costa de la Luz
Ronda
SIERRA NEVADA
Almería
Jerez de la Frontera
SIERRA DE RONDA
Málaga
Motril
Berja
Cádiz
Marbella
Costa del Sol
Gibraltar (U.K.)
Algeciras
Strait of Gibraltar
M E D I T E R R A N E A N   S E A
Cueta (Spain)

Melilla (Spain)

N
W   E
S

SPAIN

GIBRALTAR

▼ **Olive groves**
Olive trees thrive in the arid climate and dusty soil of southwest Spain and Portugal. The fruit are pressed to make oil or eaten whole.

▶ **Citrus harvest**
In spring, Seville is filled with the sweet scent of orange blossom. Seville's bitter oranges are used for making marmalade.

69

Iberian peninsula

THE THREE INDEPENDENT IBERIAN NATIONS ARE ALL democracies, although Portugal and Spain were dictatorships until the 1970s and Andorra was not fully democratic until 1993. Spain is now ruled by a king, while Portugal is a republic. Both have been members of the European Union (EU) since 1986. This has improved the standard of living, which in the past was often very poor.

The warm climate means that a wide variety of crops can be grown, including oranges, lemons, olives, melons and sunflowers. Portugal obtains cork from a kind of oak tree. Both Portugal and Spain have vineyards, and wines, sherries and ports are produced. There are large fishing fleets. The catch includes sardines, tuna, anchovies and cod. The food of the region includes spicy sausages, paella (a dish of rice mixed with seafood, chicken or ham, vegetables and garlic) and all sorts of delicious snacks known as tapas.

▲ **Spanish police**
Spanish mounted police exercise their horses. There are three police forces in Spain – the Guardia Civil, the Municipal Police and the National Police. Each has different responsibilities.

▲ **Casks of port**
Port takes its name from Porto, a town in Portugal. It is a strong, dark, sweet wine, which is traditionally drunk after dinner. It is aged in oak barrels.

▲ **Fish and fishing**
Fishing is a traditional part of the economy along the coasts of Spain and Portugal. In recent years there have been clashes with the fishing fleets of other EU countries over catch quotas.

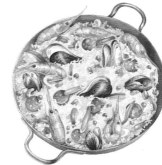

▲ **Paella**

▲ **Cork-tipped shuttlecocks**

▲ **Forests of cork**
Portuguese trees called cork oaks have a thick bark which can be harvested repeatedly to provide cork. But this traditional industry is being threatened by the introduction of plastic stoppers for wine bottles.

▶ **Follow the sun**
The Iberian peninsula attracts many tourists. Some come for sports and outdoor activities; others for the sunny beaches or the beautiful old towns.

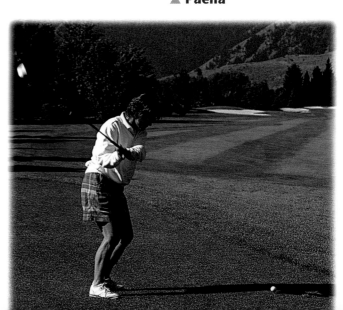

▲ **Rope-making**
Spain produces a large number of fibre crops, including cotton, flax, hemp and esparto grass. These are used in the manufacture of rope, yarn, fabrics, textiles and paper.

There are reserves of iron, coal and copper, and factories produce cars and textiles. A major industry throughout the region is tourism, with northern Europeans flocking to the sunny beaches of Spain's Costa del Sol, Portugal's Algarve, the Canary Islands and Majorca (the largest of the Balearic Islands).

About two-thirds of Portuguese are country-dwellers, but over three-quarters of the Spanish population now live in towns and cities. Many of these include historical centres with impressive castles, palaces and cathedrals as well as industrial and residential suburbs. The Spanish capital is Madrid, which is located right in the middle of the country, on the hot Meseta. Spain's second city is Barcelona, a lively cultural and commercial centre on the Mediterranean coast. Lisbon, the Portuguese capital, is a seaport at the mouth of the River Tagus.

▲ **On the beach**
Tourism has brought wealth to Spain, but has had a major impact on the Spanish environment. Hotels line the southern coasts and even remote regions are dotted with campsites.

▼ **Intrepid sailor**
This statue celebrates the pioneering Portuguese navigator, Vasco de Gama (c.1460–1524). He discovered a new route to the East round the southern tip of Africa.

◀ **City visions**
A sculpture of gold dazzles the eye in Barcelona, the chief city of the Catalonia region, Spain. Barcelona is famous as a centre of culture and the arts.

▲ **Industrial Spain**
Spain is one of Europe's major producers of cars and other vehicles, with factories in Madrid, Barcelona, Valencia and Saragossa. Vehicles are the country's most important export.

Iberian peninsula

SPANISH, OFFICIALLY BASED ON CASTILIAN, THE FORM of the language spoken in the old kingdom of Castile, is spoken throughout Spain. Portuguese is spoken throughout Portugal. Both languages have spread around the world and are spoken in Central and South America. However many other languages and dialects are still to be heard in the Iberian peninsula and its islands.

The Basques, who live in the northeast and in France, speak Euskara, which is not related to any other known language. They are a very ancient European people, with a history of seafaring. The Galicians of the northwest claim Celtic ancestry: their language, Galician, is related to Portuguese. The region of Catalonia is centred upon Barcelona, and extends into Andorra and France. It also has its own language, Catalan. Roma (gypsies) have influenced the dance and music of Spain's Andalucia region. The Canary Islanders include descendants of the Guanche, a people related to the Berbers of North Africa.

▲ Swirling skirts
The passionate flamenco is danced to the music of guitars, sometimes with castanets. It comes from around Seville and Cádiz.

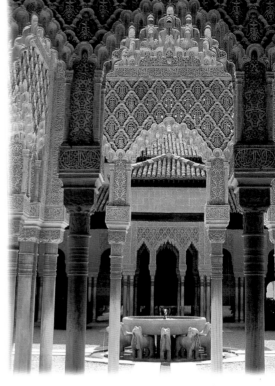

▲ The Alhambra
This royal palace and fortress was built by the Moorish rulers of Granada, Spain, between the 1000s and the 1400s. It is one of the world's most beautiful buildings.

▶ Ceramic tiles
The Iberian peninsula is famed for its fine ceramics. These tiles grace the Plaza de España in Seville, which was built in 1929 for a large exhibition.

▲ Living faith
The Iberian peninsula is staunchly Roman Catholic. Almost every town or village seems to honour its patron saint. The Holy Week before Easter is marked with festivals (fiestas) and processions.

▶ Squashed tomatoes
The town of Buñol in the eastern province of Valencia, Spain, holds an unusual festival every August. During the Tomatina, people pelt each other with tomatoes!

◀ Saints and madonnas

Images of Christian saints and of the Virgin Mary adorn every Spanish cathedral and church. Religious icons are also to be seen in the home, on postcards and even on key-rings.

▶ Salvador Dali

This Spanish artist was born at Figueras in 1904. In 1928 he joined the surrealist movement in Paris, France, painting haunting pictures with dream-like imagery.

The modern national borders of the Iberian peninsula do not match those of the peoples and cultures, and so many regions call for independence. This has led to some power being devolved from central government, but a violent separatist campaign continues in the Basque country.

The Iberian peninsula has a rich cultural history. Its writers have included Portugal's greatest poet, Luis de Camões (1524–80), Miguel de Cervantes (1547–1616), who wrote the Spanish classic *Don Quixote*, the Spanish painter Francisco de Goya (1746–1828) and the Catalan architect Antonio Gaudí (1852–1926).

▲ Spanish guitar

The region is strongly Roman Catholic, and everyday life is marked by colourful festivals in which statues of the Virgin Mary and saints are carried through the streets. Other festivals celebrate historical battles, trading fairs, horse-riding and bullfighting. Traditional costume is worn for many of these. Popular folk music includes the fiery flamenco music of Andalucia and soulful Portuguese songs known as fado.

▼ Art in Bilbão

The new Guggenheim Museum of Modern Art is located in Bilbão, in the Basque country. The gallery's titanium exterior is designed to look like rolling waves. The collection inside includes works by Spanish-born Pablo Picasso (1881–1973).

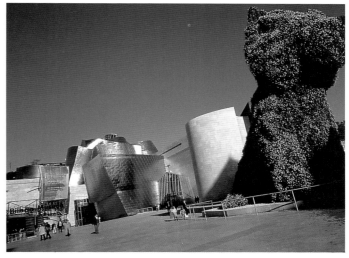
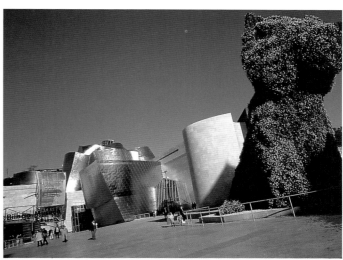

▼ Red rag to a bull

Bullfighting is a very ancient tradition in Spain and tens of thousands of bulls are still killed each year. But is it truly a sport or a cruel torment?

◀ Fairytale architecture

The best-known example of Antonio Gaudí's surreal style of architecture is the Sagrada Familia church in Barcelona. He worked on this building from 1884 until his death in 1926.

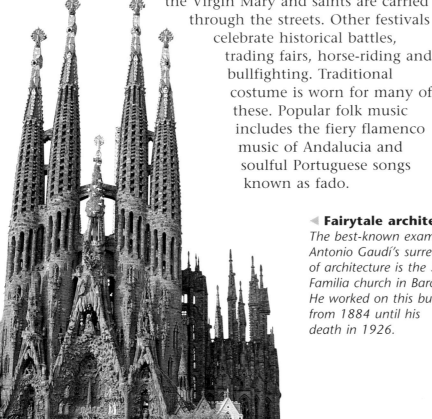

Iberian Peninsula

PREHISTORIC CAVE PAINTINGS AT ALTAMIRA, in northern Spain, date back to 12000BC. Many different peoples settled in the region in ancient times, including Iberians from North Africa and Celts from Europe. They were followed in ancient times by Phoenicians, Greeks, Romans and Jews. From about 200BC until AD475, Spain was a very important part of the Roman empire. Modern Spanish and Portuguese both developed from Latin, the language of ancient Rome.

The next invaders were Germanic peoples, including Franks and Visigoths. From AD711 onwards they were pushed back to the far north by armies of Moors – Berbers and Arabs – invading from North Africa. The lands ruled by these Muslims became known al-Andalus. Great cities with fine palaces and mosques were built. From the north, Christian armies began to fight to regain control, but this was not completed until 1492.

▲ **Islamic detail**
The Arabic script on this Spanish stonework shows the Moorish legacy stamped on to Spanish architecture during the Muslim occupation in medieval times.

▲**Cave paintings**
The prehistoric artists of Altamira painted the bulls, boars and bison that they hunted. For colour, they used red ochre from the soil and black soot.

▶ **Philip II**
Born in 1527, Philip II became one of the most powerful and ruthless Spanish kings.

▼ **Ferdinand Magellan**
This Portuguese seafarer, in Spanish service, set sail for the Americas in 1519. He sailed on to the Pacific, but was killed in the Philippines in 1521. His ship returned to Spain, becoming the first to sail around the world.

▲ **The heroic knight**
Rodrigo Díaz was born at Burgos in about 1043. He was a Christian knight who mostly fought against the Moors, although sometimes he fought for them. He captured Valencia in 1094. He became known as El Cid (from the Moorish Sidi, or 'lord').

▲ **The rebel**
Spain lost its huge South American empire in the 1800s. Argentinean-born José de San Martín (1778–1850) helped to liberate his native country as well as Chile and Peru.

LISABONA

◄ Lisbon, 1755
In the 1700s, the Portuguese capital was a wealthy city with many fine buildings. In November 1755 it was destroyed by a terrible earthquake just out to sea, which caused floods and fires. Ten thousand people died.

The patchwork of small Catholic kingdoms gradually joined together during the Middle Ages. Spain was united in 1512, and it ruled Portugal too from 1586 to 1646. Seafarers from both countries led European exploration and colonization of the Americas and traded with Africa and the Far East. They looted, fought and traded around the world and became very wealthy and powerful.

However, wars with other European nations weakened both countries. In the 1800s their American colonies mostly broke away from European rule. There was great political strife in Spain, which was torn apart by a bitter civil war between 1936 and 1939. A fascist dictator, General Franco (1892–1975) ruled Spain until 1975, when it became a kingdom once again. Portugal too was ruled by dictators for 50 years, from 1926 onwards.

Portugal gave up its remaining colonies in Africa, and the region as a whole became seen once again as being central to the development of Western Europe, rather than as a backwater.

▲ The general
In 1936 Francisco Franco launched an attack on Spain's government. After three years of civil war, assisted by Nazi Germany, he came to power in 1939.

► Democratic Portugal
After long years of dictatorship, Portugal celebrated its return to democracy in 1976.

AD	TIMELINE
264	Germanic invasions
711	First Moorish invasion
1085	Christians capture Toledo
1086	Second Moorish (Almoravid) invasion
1143	Portugal an independent kingdom
1147	Third Moorish (Almohad) invasion
1212	Moors defeated at Las Navas de Tolosa
1278	Status of Andorra confirmed
1440s	Portugal expands overseas
1479	Kingdoms of Castile and Aragon unite
1492	Fall of Granada, last Moorish kingdom; Columbus discovers the Americas
1516	Carlos I, of the Austrian Habsburg dynasty, becomes king of Spain
1586	Portugal under Spanish rule (until 1646)
1701	War of Spanish Succession (until 1713)
1713	Gibraltar becomes British
1807	French begin occupation of Iberia (until 1814)
1919	Conflict with anarchists in Barcelona
1926	Dictatorship in Portugal
1936	Spanish Civil War (until 1939, when Franco becomes Spanish dictator)
1975	Spain becomes democratic monarchy
1976	Portugal returns to civilian rule
1982	Increased terrorist activity by ETA (demanding independence for Basque country); ceasefire 1998
1986	Spain and Portugal join EEC (later EU)

Italy and Malta

The Italian peninsula is a long, narrow strip of land stretching southwards from western Europe into the Mediterranean Sea. In the north, it borders the icy peaks, glaciers and blue lakes of the Alps. These drop to a low-lying plain around the river Po, a wide open patchwork of fields and towns which are fringed by coastal marshes and lagoons around the city of Venice. Flat lands also border Slovenia, to the east of the Dolomite range.

▲ **The eternal city**
The capital of modern Italy, Rome was the centre of an ancient empire and of Christian Europe in the Middle Ages.

The rocky ridge of the Apennine Mountains runs down the spine of the peninsula, with fertile, vine-covered slopes descending to the coast. The rivers Tiber and Arno rise in these peaks. In the far south are dusty plains and scrub-covered hills. This is a danger zone for earthquakes and volcanic eruptions.

▲ **Mt Etna**

The central Mediterranean is divided into three smaller seas – the Ligurian and Tyrrhenian to the west of Italy and the Adriatic to the east. Offshore lie a number of small islands, such as Elba, Ischia, Capri and the Lipari Islands. There are two large islands as well, Sardinia and Sicily. Far to the south, lying between Sicily and the North African coast, are the islands which make up the small independent nation of Malta.

▼ **The smallest mammal**
The tiny Etruscan shrew weighs less than a sugar lump.

▼ **Tuscan landscape**
Tuscany is a region of central Italy. It includes medieval hilltop towns, mountain slopes and coastal plains, farmland and vineyards.

FACTS

ITALY
Repubblica Italiana
Area: *301,268 sq km*
Population: *57.4 million*
Capital: *Rome*
Other cities: *Milan, Naples, Turin*
Highest point: *Monte Bianco (Mont Blanc) (4,807 m)*
Official language: *Italian*
Currency: *Italian lira*

MALTA
Repubblika ta' Malta
Area: *316 sq km*
Population: *0.4 million*
Capital: *Valletta*
Official languages: *Maltese, English*
Currency: *Maltese lira*

SAN MARINO
Repubblica di San Marino
Area: *61 sq km*
Population: *0.03 million*
Capital: *San Marino*
Official language: *Italian*
Currency: *Italian lira*

VATICAN CITY
Stato della Cittá del Vaticano
Area: *0.44 sq km*
Population: *1,000*
Capital: *Vatican City*
Official languages: *Latin, Italian*
Currency: *Italian lira*

▲ By the seaside
The fishing town of Positano lies to the west of Amalfi. Its buildings cling to the steep, rocky coast.

▲ Bay tree

▼ Subterranean dream-world
In the eastern foothills of the Appennines lies a labyrinth of caves, stretching back for 13 kilometres. They include the 'Room of Candles', with its stunning displays of stalactites and stalagmites.

▼ The Maltese capital
Since the 1500s, sailors entering the natural harbours of Valetta have been greeted by the dome of its cathedral.

SAN MARINO

VATICAN CITY

ITALY

MALTA

MALTA

Italy and Malta

THE REPUBLIC OF ITALY OCCUPIES ALMOST ALL THE ITALIAN peninsula as well as Sardinia and Sicily. Its capital is the ancient city of Rome, on the river Tiber.

Italy is a democratic country. It was the Treaty of Rome, signed in 1957, which set up the European Economic Community (today's European Union, or EU).

The wealthiest and most industrialized part of the country is the north, which is linked to the rest of Western Europe by road and rail tunnels through the Alps. The south of the country remains poor. Over the years many of its workers have emigrated to northern Europe, the United States, Argentina or Australia.

Nearly 70 percent of Italians are now city-dwellers, but agriculture is still very important for the economy. Farms grow wheat, rice, maize, olives and tomatoes. The vineyards of the Tuscany region are famous for their wine. Indeed, Italy as a whole is the world's largest wine producer. Italian food, much of it based on pasta, risotto, polenta or pizza dishes, has become a worldwide favourite.

▲ Pizza
Pizza has a delicious topping of tomato sauce, cheese and chopped vegetables, meats or fish.

▲ Gelati means ice cream
Italians are the world's most famous makers of ice cream. Many emigrants took these skills with them to northern Europe and to the USA.

◄ The streets of Rome
In his smart uniform, a traffic policeman directs cars on the busy Roman roads.

► Venice regatta
Venice, built around canals and lagoons, became a great power in the Middle Ages. Its history is celebrated each year with a series of costumed regattas.

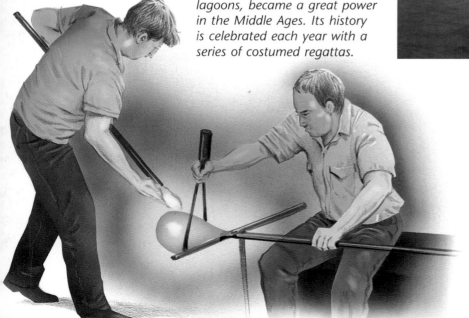

◄ The glass-blower
Fine glass has been made on the Venetian island of Murano since 1291, when the glassworks were moved from the city itself owing to the risk of fire. The island still produces glassware for the table, as well as necklaces, other jewellery and ornaments.

◄ **City of style**
Milan is the fashion
capital of Italy. Famous
fashion houses include
Gucci, Prada, Armani
and Valentino.

Italy has few mineral resources, but has developed many successful industries, including plastics, textiles, leather goods, fashion, cars, computers and household goods. Tourism in Italy dates back over 500 years. The country offers visitors winter sports or sunny beaches as well as ancient sites, beautiful old cities and towns.

▲ **Ferrari fashion**
First designed by Italian racing driver Enzo Ferrari (1898–1988), Ferraris are admired for their speed and beauty. These luxury sports cars are exported around the world.

Set within Italy are two small patches of independent territory. The republic of San Marino, in the Apennines, was recognised as independent in 1631. It attracts many tourists. Vatican City, the world's smallest state, is part of the city of Rome. It serves as the world headquarters of the Roman Catholic Church.

Malta lies far to the south of Italy, on the ancient shipping lanes that criss-cross the Mediterranean. It is a democratic republic and its economy depends on tourism and ship repairs.

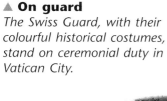

▲ **On guard**
The Swiss Guard, with their colourful historical costumes, stand on ceremonial duty in Vatican City.

▲ **The Polish pope**
John Paul II became pope in 1978. He was the first non-Italian to hold the office in 450 years.

▲ **Maltese holidays**
Sunshine lures many tourists from northern Europe to the islands of Malta and Gozo. Many outsiders buy holiday or retirement homes there, too.

► **Dock work**
Once servicing ships of the British navy, Malta's docks now repair all kinds of merchant shipping.

Italy and Malta

ITALIAN, DIRECTLY DESCENDED FROM THE LATIN language of the ancient Roman empire, is spoken all over Italy. The dialect spoken on the island of Sardinia is said to be the closest to the original Latin. In the far north there are a few speakers of French, German and Slovenian. In the high valleys of the Dolomites, a language called Ladin can also be heard. It is related to the Romansch language of Switzerland.

Today's Italians are descended from all sorts of native peoples, settlers and invaders. These included ancient Italian tribes, Etruscans, Romans, Greeks, Celts, Goths, Vandals, Lombards, Sards and Normans. The descendants of these peoples mostly merged together over the centuries. Today, the Italians' sense of identity depends more on the city or region in which they live than on their ancestry.

▲ Pavarotti
Italian is the language of opera. Luciano Pavarotti brought classical music to a wider audience when he formed the 'Three Tenors' with Placido Domingo and José Carreras.

▲ Pasta making
Pasta is made from flour and eggs and may be coloured with spinach or tomato. It comes in many shapes and sizes.

▶ St Francis of Assissi
Giovanni Bernadone (c.1181–1226) was the son of a rich merchant. He was nicknamed Francesco. In 1205 he gave up his riches to become a monk, working with the poor and sick. He preached simplicity and a love for all living creatures. Thousands followed his teachings and joined the Franciscan order of monks.

◀ Carnival time
Of all the world's carnivals, the one staged in Venice before the Christian period of Lent is the most elegant. Masked balls are held in the costume of the 1700s. In those days, carnival went on for two months!

▶ Football magic
Italian football clubs are world beaters, attracting players from all over the world. Here, the Brazilian Ronaldo takes the field for Inter Milan.

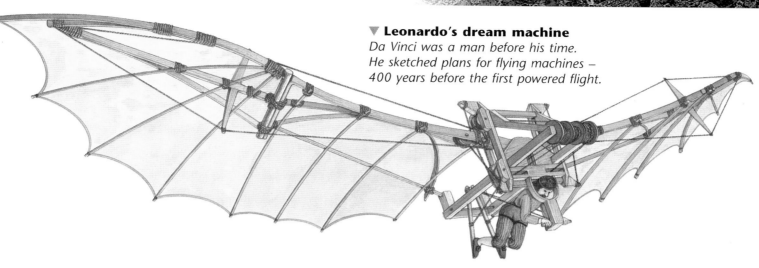

▼ Leonardo's dream machine
Da Vinci was a man before his time. He sketched plans for flying machines – 400 years before the first powered flight.

The traditional way of life in Italy has changed a little in recent years, but it is still very strongly influenced by the Roman Catholic Church and the family. Many colourful regional festivals mark Christian saints' days or Holy Week.

Italy has probably contributed more to European civilization than any other country. Ancient Rome, heavily influenced by ancient Greece, was a centre of literature and the arts, of philosophy and law-making, of engineering and technology. In later phases, the Italian city states produced great poets such as Dante (1265–1321), artists such as Michelangelo (1475–1564) and Leonardo da Vinci (1452–1519) and musical composers such as Monteverdi (1567–1643). Opera is still very popular today, arousing as much passion as football.

The Maltese way of life has been strongly influenced by Italian traditions. The Maltese themselves are descended from the various peoples who, at one time or another in history passed through their islands – including Greeks, Phoenicians, Arabs, Normans and English, as well as Italians.

▲ Maltese cross
The Maltese cross is the badge of the Knights of St John, a religious fighting order that used Malta as its base from 1530 to 1798.

▲ Dante Alighieri
Born in 1265, Italy's greatest medieval poet wrote the Divine Comedy, *a vision of heaven and hell. It was the first work of genius to be written in the Italian language, rather than Latin.*

▶ Measured time
This clockface towers above St Mark's Square, in Venice. Two bronze figures strike the hour and, every January 6th, an angel and the Three Kings appear. The clock was completed in 1506.

◀ Red Bologna
From the end of World War II in 1945, communists played an important role in Italian politics. Their power base was the northern city of Bologna.

Italy and Malta

THE CITY OF ROME IS SAID TO HAVE been founded in 753BC, at a time when the Etruscans ruled over central Italy and Greek colonies were being founded in the south. Rome overthrew Etruscan rule in 509BC.

In the following centuries, the Romans conquered the rest of Italy and defeated a powerful rival, the North African city of Carthage. They founded a great empire, which stretched from Spain to Syria. By AD476 the western half of this empire, including Italy, had been overthrown by Germanic invaders.

For the next 1,400 years Italy remained divided. There were invasions by Lombards and Byzantines. Normans carved out kingdoms for themselves in the south. The north fell under the rule of the Holy Roman Emperors, from across the Alps. However the city of Rome, headquarters of the Roman Catholic Church, was still powerful as the centre of the Christian world.

▲ Roman artefacts
Archaeological finds of pottery, glass, coins and weapons have helped us understand what everyday life was like during the days of the Roman empire.

▲ Ancient stones
Megalithic (great stone) monuments mark Stone Age burial sites on the island of Malta. Some date back to about 3500BC.

▼ Pompeii destroyed
In AD79 Vesuvius, a volcano on the Bay of Naples, erupted. Nearby towns, such as Pompeii and Herculaneum, were buried in ash and mud. Thousands were killed. In the 1700s these ancient towns were excavated, revealing streets, houses and gardens.

▼ The Colosseum
This amphitheatre was the biggest in the Roman empire. It was opened by the Emperor Titus in AD80. Many ancient monuments still stand in the city of Rome today.

▼ Trajan's Column
Completed in AD113, this marble monument records the triumphs of the Roman Emperor Trajan (AD53–117).

▶ Roman coins
Long before a single European currency was launched in 1999, most of Europe used the same coins – those of the Roman empire.

▼ In the arena
The Romans loved cruel spectacle. In arenas such as this, crowds watched armed fighters called gladiators kill each other, and roared as slaves were torn apart by wild animals.

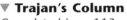

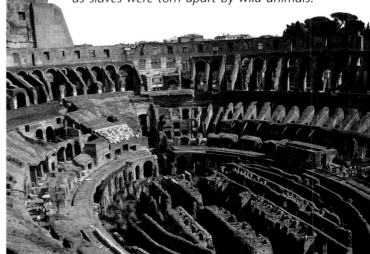

◀ **Catherine de' Medici**
The powerful and ruthless Medici family ruled Florence and, later, Tuscany, from the 1400s to the 1700s. Catherine de' Medici (1519–89) married Henry II, king of France.

▶ **Garibaldi**
In 1861 a patriotic adventurer called Giuseppe Garibaldi landed in Sicily and began a military campaign to reunite his divided country. He and his 1,000 'Red Shirts' experienced setbacks and defeats, but Garibaldi lived to see union in 1871.

AD	TIMELINE
117	Roman empire at greatest extent
568	Lombards invade northern Italy
917	Arabs capture Sicily
1090s	Normans invade Malta, Sicily
1450	Height of the Renaissance
1494	France invades Italy
1530	Knights of St John (Malta) fight Turks
1631	San Marino independent
1798	France takes Malta
1800	French kingdom of Italy (until 1815); Britain takes Malta
1848	Revolution fails to end Austrian rule
1859	Austria loses most of Italian possessions
1860	Garibaldi's patriots liberate Sicily
1871	Unification of Italy
1915	Italy joins Allies in World War I (until 1918) against Central powers
1922	Fascist leader Mussolini seizes power
1940	Italy joins World War II on German side; changes sides 1943; Mussolini shot
1957	Italy founder member of EEC (later EU)
1964	Malta independent from British rule

A number of republics and cities grew up in the north during the Middle Ages. These became centres of commerce, and it was here that learning and the arts flourished during the period of cultural rebirth known as the Renaissance.

Italy experienced periods of rule by the French, Spanish and Austrians. During the 1800s a movement grew up to reunite the country, under the leadership of Giuseppe Garibaldi (1807–82). By 1871 Italy was a united kingdom.

Having supported the Allies in World War I (from 1915 until 1918), Italy soon came under the rule of a fascist dictator called Benito Mussolini. Disastrously, Italy was allied with Germany during World War II and in 1945 Mussolini was shot dead by rebel fighters.

▶ **Father of fascism**
The fasces *were an emblem of state authority in ancient Rome. The name was adopted by the Fascists, a group of right-wing nationalists. Their leader, Benito Mussolini, marched on Rome and seized power in 1922.*

In the years after war, Italy was rebuilt. It enjoys democratic government, despite problems with organized crime.

▶ **Mafia on trial**
The years after World War II saw the rapid rise of the Mafia, a murderous criminal organization based in Sicily. They were protected by corrupt police and politicians.

83

Central Europe and the Baltic

The northern part of Central Europe borders the Baltic Sea. Its eastern coast is low-lying, bordering sand dunes, bogs, heaths and forests. Estonia and Latvia lie on the Gulfs of Finland and Riga, with Lithuania's short coastline stretching south to Kaliningrad, a small corner of Russian territory. Below the Gulf of Gdansk, the Baltic coast continues westwards to Germany.

▲ **Prienai, Lithuania**
Only one in three Lithuanians are now country-dwellers. Farmers grow potatoes, beet and rye and raise pigs and chickens.

Most of Central Europe is far from the sea. The climate may be warm and sunny in summer, but the winters are long and cold. Much of Poland is a great plain, which stretches eastwards into Belarus and Russia. Parts are heavily forested or dotted with lakes, while others are farmed or industrialized. In southern Poland the land rises to mountains along the borders of the Czech Republic and the Slovak Republic.

The Sudetes mountain range runs along the northern Czech border, while to the southeast the Bohemian Forest stretches into Germany. The Czech lands include wooded hills and rolling farmland, drained by the rivers Vltava, Elbe and Morava. The Slovak Republic is more mountainous, with the Tatra range descending to fertile lowlands along the river Danube. Northern Hungary is a land of hills and low mountains, while the southeast forms a wide plain, with a rich, black soil ideal for farming.

▼ **In the Tatra**
The mountains between Poland and the Slovak Republic are home to animals such as the Tatra chamois, an endangered species, and the European suslik.

FACTS

CZECH REPUBLIC
Ceská Republika
Area: *78,864 sq km*
Population: *10.3 million*
Capital: *Prague*
Other cities: *Brno, Ostrava*
Highest point: *Snezka (1,602 m)*
Official language: *Czech*
Currency: *Koruna*

ESTONIA
Eesti Vabariik
Area: *45,100 sq km*
Population: *1.5 million*
Capital: *Tallinn*
Other cities: *Tartu, Kohtla-Järve*
Highest point: *Munamagi (318 m)*
Official language: *Estonian*
Currency: *Kroon*

HUNGARY
Magyar Köztársaság
Area: *93,032 sq km*
Population: *10.2 million*
Capital: *Budapest*
Other cities: *Debrecen, Miskolc*
Highest point: *Kekes (1,015 m)*
Official language: *Magyar*
Currency: *Forint*

LATVIA
Latvija – Latvijas Republika
Area: *64,600 sq km*
Population: *2.5 million*
Capital: *Riga*
Other cities: *Jelgava, Daugavpils*
Highest point: *Jaizina (311 m)*
Official language: *Latvian*
Currency: *Lats*

▲ Lake Balaton
Central Europe's largest lake lies surrounded by orchards and farms to the southwest of Budapest, in Hungary's wine-producing region. In winter, its shallow waters freeze over.

LITHUANIA

ESTONIA

LATVIA

▼ In Bohemia
Northern Bohemia, in the Czech Republic, is a region of woods and pretty old towns and villages. A day's hiking may be followed by a meal at a restaurant or a bottle of the famous Czech beer, Pilsener.

POLAND

CZECH REPUBLIC

SLOVAK REPUBLIC

HUNGARY

Map labels

Estonia: Hiumaa, Saaremaa, Tallinn, Kohtla-Järve, Lake Peipus, Parnu, Tartu, Munamagi 318 m, Gulf of Riga

Latvia: Ventspils, Jurmala, Saldus, Riga, Jelgava, Liepāja, Gaizina 311 m, Daugavpils, Siauliai

Lithuania: Panevezys, Klaipeda, Utena, Lithuania, Ukmerge, Nemunas (Neman), Kaunas, Vilnius 311 m, Kaliningrad (RUSSIA), Kaliningrad

Russia

Belarus

Gulf of Gdansk, Gdynia, Gdansk, Kolobrzeg, Elblag, Olsztyn, Szczecin, NORTH EUROPEAN PLAIN, Bydgoscz, Bialystock, Gorzow Wielkopolski, Torun, Poznan, Plock, Kalisz, Warsaw, POLAND, Glogow, Lodz, Radom, Lublin, Odra (Oder), Wroclaw, Kielce, Chelm, Walbrzych, Czestochowa, SUDETES MOUNTAINS, Bytom, Karlovy Vary, Prague, Katowice, Krakow, Rzeszow, Plzen, Pardubice, Tychy, Tarnow, BOHEMIA, CZECH REPUBLIC, Ostrava, Bielsko-Biala, CARPATHIAN MOUNTAINS, Olomouc, MORAVIA, Zilina, Presov, Cesky Budejovice, Brno, Rysy Peak 2,499m, Kosice, Trencin, GERMANY, UKRAINE, AUSTRIA, SLOVAK REPUBLIC, Miskolc, Nytra, Bratislava, Mt. Kekes 1,015m, Debrecen, Danube, Györ, Budapest, Tatabanya, Koros, Szombathely, HUNGARY, Bekescsaba, Lake Balaton, Tisza, ROMANIA, Kaposvar, Szeged, Pécs, CROATIA, YUGOSLAVIA

▲ European bison

◄ Czech crags
Eroded into a fantstic pillar, Pravicky Cone can be seen in northern Bohemia, a mountainous region of the Czech Republic.

◄ Budapest
Hungary's Houses of Parliament lie at the centre of the capital, Budapest, on the banks of the river Danube.

85

Central Europe & the Baltic

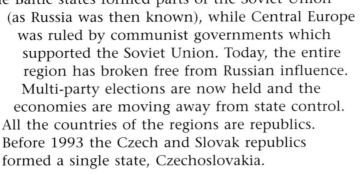

FROM THE 1940s UNTIL THE YEARS 1989–91, the Baltic states formed parts of the Soviet Union (as Russia was then known), while Central Europe was ruled by communist governments which supported the Soviet Union. Today, the entire region has broken free from Russian influence. Multi-party elections are now held and the economies are moving away from state control. All the countries of the regions are republics. Before 1993 the Czech and Slovak republics formed a single state, Czechoslovakia.

The Baltic states lack mineral resources and need imports to support their industries. They produce chemicals, textiles and heavy machinery. The forests provide timber for making paper and matches. Many factories and mills are old-fashioned. Farms grow staple crops and raise cattle and pigs.

Poland has reserves of coal, sulphur, copper, silver, lead, salt and natural gas. Shipbuilding and repair make up a major industry on the Baltic coast and factories produce vehicles, heavy machinery and footwear. Wheat and rye are grown and potatoes are a major crop.

▲ Pig-farming
Many farms in the Baltic states raise pigs. Bacon and dairy products are major exports in Lithuania and Estonia.

▲ Staple crops
In Central Europe, staple crops – those that provide the basis of people's everyday diet – include wheat and rye, for making bread, and potatoes.

▶ Matchmakers
The conifer forests of Central Europe and the Baltic states provide match wood as well as pulp for paper.

◀ On the march
These Lithuanian soldiers are on parade to mark Independence Day. The country became an independent state in 1991.

▼ Shipbuilding
Ship repair and construction is a traditional heavy industry around the Baltic coast, especially in the Polish port of Gdansk.

FACTS

LITHUANIA
Lietuva, Lietuvos Respublika
Area: *65,200 sq km*
Population: *3.7 million*
Capital: *Vilnius*
Other cities: *Kaunas, Klaipeda*
Highest point: *Juozapines (292 m)*
Official language: *Lithuanian*
Currency: *Litas*

POLAND
Rzeczpospolita Polska
Area: *323,250 sq km*
Population: *38.6 million*
Capital: *Warsaw*
Other cities: *Lodz, Krakow, Wroclaw*
Highest point: *Gerlachovsky Stit (2,656 m)*
Official language: *Polish*
Currency: *Zloty*

SLOVAK REPUBLIC
Slovenská Republika
Area: *49,012 sq km*
Population: *5.3 million*
Capital: *Bratislava*
Other cities: *Kosice,*
Highest point: *Rysy Peak (2,499 m)*
Official language: *Slovak*
Currency: *Koruna*

◄ Old Prague
The Czech capital has many fine old buildings dating from the 1300s and 1400s, when the city was the centre of the Bohemian kingdom.

▲ Cherry ripe
The orchards of Hungary produce delicious cherries, for canning and jam-making.

When Czechoslovakia split in two, much of the industry lay on the Czech side of the border. It included glass-making, iron and steel. Slovakia depends more on its farmland. Hungary has reserves of bauxite, coal and gas and produces chemicals, plastics, aluminium, vehicles and electrical goods. Its orchards grow cherries and other fruits for making jams.

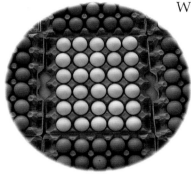

▲ Dyed eggs
Onion skins are used to dye the shells.

Central Europe is famous for its delicious food, such as Hungarian goulash (a stew made of meat flavoured with sour cream and paprika) and Polish *pierogi* (dumplings stuffed with mushrooms and meat). Alcoholic drinks include Polish vodkas, Czech beers and strong red wines from Hungary.

Tourism is a growing industry in Central Europe. Major attractions include the old cities of Krakow, Prague and Budapest, the Tatra Mountains and the hot springs and health resorts of the Czech Republic.

◄ The taste of paprika
The red fruit of a type of capsicum plant is dried and then ground to make paprika. This peppery spice is used in dishes such as Hungarian goulash.

▲ Visiting Central Europe
The tourist industry is growing rapidly in Poland, Hungary and the Czech Republic. Here, visitors admire a waterfall at Souteska, northern Bohemia.

► Hungarian spas
Warm springs near Miskolc, in the north of Hungary, offer a chance to relax with friends. The water can be as hot as 30°C. Rich in minerals, it is said to cure many illnesses.

Central Europe and the Baltic

PEOPLES LIVING ON THE EAST COAST OF THE BALTIC Sea include the Estonians, Latvians and Lithuanians, each of whom have their own language. The Estonian language is related to Finnish and Hungarian. During the years that this region formed part of the Soviet Union, many Russians, Ukrainians and Belarussians also settled here.

The Poles, Czechs and Slovaks are all Slavic peoples, speaking languages which are separate but related to each other. The Magyar people, who invaded southern Central Europe about 1,200 years ago, make up 90 percent of the Hungarian population. Hungary is also home to small numbers of Germans and neighbouring peoples such as Czechs, Serbs, Croats and Romanians.

Central Europe as a whole has a large Roma (gypsy) population, which has suffered from racism and injustice in many areas over the years. Many Roma and most of the region's large Jewish populations were scattered or murdered in terrible death camps during the occupation of the region by Nazi Germany during World War II (1939–45).

▲ Mixed nationality
Ceramic tiles adorn this chemist's shop in the old border town of Sopron. Allotted to Austria after World War I, Sopron voted to join Hungary in 1924.

▲ Horsemanship
A love of horses and riding was part of the Magyar way of life long before they settled in Hungary, and remains a tradition today.

▼ Still on the road?
The traditional painted caravans of the Roma are a rare sight these days. In Central Europe, only ten percent of the Roma people are still travellers.

▼ Dressed for a feast
These Poles wear traditional dress. Such outfits are still worn for folk dances and festivals. Bright colours and intricate embroidery are a feature of all the national costumes of Central Europe.

▲ Eggshell paint
Hand-painted eggshells go on display in Budapest. Egg painting is an Easter tradition in Central and Eastern Europe, where it has become an art form.

▼ A long history
The Hungarian National Museum celebrates a history that goes back to the 800s, when a legendary Magyar chief called Arpad conquered the region and set up a powerful kingdom.

Catholic Poland

Churches have been an integral part of the Polish landscape since AD966, when the country became Christian. Ninety-five percent of Poles are Roman Catholics, and Pope John Paul II is Polish.

Plucked strings

The zither is an instrument used in Central European folk music. It has a flat soundbox and its jangling strings may be plucked with the fingers or a plectrum.

Central Europe is mostly Christian, with the Roman Catholic Church especially strong in Poland. The monastery of Jasna Góra at Czestochowa attracts many Catholic pilgrims to honour an icon (or holy picture) known as the Black Madonna.

The region has contributed greatly to European culture. The musical composer Frédéric Chopin (1810–49) was Polish, while composer Antonín Dvořák (1841–1904) was a Czech. Classical music has been much influenced by regional folk music and dances. It was two Poles who completely changed our understanding of science. The astronomer Nicolaus Copernicus (1473–1543) described how the Earth travels round the Sun. The great physicist Marie Curie (1867–1934) researched into radioactivity.

Antonin Dvořák (1841–1904)

Dvořák (pronounced D-vor-zhak) was born near Prague. A great classical composer, he was inspired by Slavonic folk music. He wrote symphonies, operas, chamber and piano music. It was in the United States that he wrote his most famous symphony, From the New World.

A new theory

Nicolaus Copernicus was a Polish astronomer who died in 1543. He realized that the Earth travels around the Sun, rather than the other way round.

Marie Curie

Marie Sklodowska (1867– 1934), shown left, was a Polish physicist who isolated the elements polonium and radium. She married the French physicist, Pierre Curie.

Central Europe and the Baltic

THE FLAT LANDS LYING BETWEEN western and eastern Europe are hard to defend. Throughout history, they have been invaded from both the east and the west. Poland has probably changed its position on the map more times than any other country in Europe.

About 2,700 years ago, the Czech and Slovak region and part of Hungary formed a major centre of the Celtic civilization. This was displaced by Germanic tribes, but in the end it was Slavic peoples who controlled most of Central Europe from the Baltic Sea to the river Danube. In about AD800 the plains of Hungary were seized by the Magyars, fierce horseback warriors from the East.

In the Middle Ages, powerful kingdoms grew up in the region, such as Poland (which united with Lithuania in 1382), Bohemia (with its capital at Prague) and Hungary. Central Europe came under attack from the east by Tartars and Turks, while from the west Germany's Teutonic Knights rode into Poland and the Baltic. The Austrian empire spread eastwards, to take in Bohemia by the 1500s and Hungary by the 1700s. Poland and the Baltic states were torn apart and invaded by the great powers of the day – Sweden, Russia, Austria and the north German state of Prussia.

▲ **The Magyars**
In about AD800 hordes of nomadic warriors called Magyars poured across the Ural Mountains from Russia and conquered the plains of Hungary.

▲ **On guard**
A soldier stands on guard outside Prague's castle, which was once the centre of a powerful medieval kingdom, Bohemia.

▲ **A stronghold**
The Lithuanian city of Kaunas was founded in medieval times and has a stormy history of warfare and destruction.

◀ **Out of the window**
An incident in Prague, in 1618, marked the start of the Thirty Years War. The envoys of the Catholic Holy Roman Emperor were thrown out of the window by the Bohemian nobles, who were Protestant. They chose instead to be ruled by a German Protestant prince. Two years later, Catholic forces completely destroyed the Bohemian forces at the Battle of the White Mountain.

Workers united
In 1980 a trade union known as Solidarity was formed in Poland. It challenged the authority of the communist government with a series of strikes in the Gdansk shipyards. Solidarity's leader, Lech Walesa, became Polish president in 1990.

After World War I (1914–18), the nations of Central Europe were independent once again. However German invasion during World War II (1939–45) brought terror and devastation to the region. Central Europe was liberated by local freedom fighters and by the army of the Soviet Union.

Post-war governments in Central Europe were effectively controlled by the Soviet Union. There were uprisings against them, but in the end it was the collapse of the Soviet Union in 1991 which spelt the start of a new age for Central Europe and the Baltic states.

AD	TIMELINE
500s	Slavs settle Central Europe
800s	Magyars seize Hungary
1038	Death of Stephen I; united Hungary
1237	Tartar invasions (until 1242)
1308	Teutonic Knights take Gdansk
1410	Poles defeat Teutonic Knights
1569	Union of Lublin – Poland, Prussia, Lithuania and Livonia (Latvia) form commonwealth
1618	Thirty Years War (until 1648) starts with Protestant revolt in Bohemia
1648	Russia and Sweden seize Polish territory
1699	Hungary under Austrian rule
1772	Poland partitioned between Russia, Austria, Prussia
1848	Hungary revolts against Austrian rule
1914	World War I (until 1918): Central powers (Germany, Austria) invade Poland
1918	Lithuania, Poland, Hungary, Czechoslovakia independent
1920	Latvia independent
1921	Estonia independent
1939	World War II (until 1945): Soviet Union annexes Baltic states 1940; Germany invades Central Europe and Baltic; massacre of Jews; Soviet Union pushes back German forces
1948	Pro-Soviet communists take power in Central Europe
1956	Anti-Soviet uprising in Hungary
1968	Czech reforms crushed by Soviet Union
1990	Democratic elections in Central Europe
1991	Baltic states recognised as independent
1993	Czech and Slovak states separate

Prague, 1968
Soviet Tanks rolled into Czechoslovakia in 1964 after the Czech government tried to introduce political reforms.

Independent Slovakia
This monument at Bankska Bystrica celebrates Slovak independence, gained when the country was part of Czechoslovakia. Today, it is independent in its own right.

The Balkans

▲ Pelicans
Rare Dalmatian and White pelicans breed in Balkan wetland regions.

The Balkan peninsula stretches southwards into the Mediterranean Sea from Central and Eastern Europe. It is at its broadest in the north, where it stretches all the way from the islands of the northern Adriatic coast to the marshy delta of the river Danube, on the Black Sea. It narrows rapidly between the Ionian and Aegean Seas, where it breaks into the ragged headlands and island chains of Greece.

The only large plains lie in the north. The region as a whole is very mountainous, taking in the Carpathian ranges, the Transylvanian Alps, the Dinaric Alps, and the Balkan, Rhodope and Pindus ranges. Even the largest of the Greek islands, Crete, is topped by a 2,456-metre peak called Mount Ida.

Winters can be cold and snowy in the northern Balkans, but are normally milder in the south. Summers are dry and hot, often fiercely so in Greece. Northern forests give way to scrub-covered rock and olive groves in the south. The Balkans lie in an extremely active earthquake zone.

The Balkans still include large areas of remote forest and mountain habitat, where wolves, brown bears and wild boar live. The warmer regions have a wide variety of snakes and lizards. Many of the region's species are threatened by loss of habitat as forests are cleared and tourist resorts are built.

▲ Slovenian workers
Slovenia is close to Western European markets for goods. It has escaped the worst of the region's strife in recent years.

◀ Mt Triglav
The Slovenian peak of Triglav (2,863 metres) rises dramatically 65 kilometres to the northwest of Ljublana. It is part of the Julian Alps range, on the border with Austria.

FACTS

ALBANIA
Republika e Shqipërisë
Area: 28,748 sq km
Population: 3.3 million
Capital: Tiranë
Other cities: Durrës, Shköder, Elbasan
Highest point: Korabit (2,751 m)
Official language: Albanian
Currency: Lek

BOSNIA-HERZEGOVINA
Republika Bosnia i Hercegovina
Area: 51,129 sq km
Population: 4.0 million
Capital: Sarajevo
Other cities: Banja Luka, Mostar
Highest point: Maglic (2,386 m)
Official language: Serbo-Croat
Currency: Bosnian dinar

BULGARIA
Republika Bulgaria
Area: 110,912 sq km
Population: 8.4 million
Capital: Sofia
Other cities: Plovdiv, Varna, Burgas
Highest point: Musala (2,925 m)
Official language: Bulgarian
Currency: Lev

CROATIA
Republika Hrvatska
Area: 88,117 sq km
Population: 4.8 million
Capital: Zagreb
Other cities: Split, Rijeka
Highest point: Troglav (1,913 m)
Official language: Serbo-Croat
Currency: Croatian dinar

GREECE
Elliniki Dimokratia
Area: 131,957 sq km
Population: 10.5 million
Capital: Athens
Other cities: Thessaloníki, Lárisa
Highest point: Olympus (2,917 m)
Official language: Greek
Currency: Drachma

▲ Hermann's tortoise

▶ Capital city
Bucharest (Bucuresti) lies on the river Dambovit. It became the capital of Wallachia in 1698 and the capital of Romania in 1861.

▲ On the Kika River
Croatia, lying between the Hungarian border and the Adriatic coast, is a beautiful country of mountains, rivers and waterfalls.

YUGOSLAVIA

CROATIA

SLOVENIA

ROMANIA

BULGARIA

BOSNIA-
HERZEGOVINA

MACEDONIA

▲ A cave-dweller
The pale, blind olm is an endangered species of salamander that is found in underground lakes and streams in Slovenia and Yugoslavia.

ALBANIA

GREECE

▲ The eastern Adriatic
This beautiful Mediterranean shoreline attracted many visitors in the 1970s and 80s. However, fighting in the 1990s destroyed the tourist industry.

◀ Nearer to God
These sheer pillars of rock in Thessaly, Greece, are known as Méteora. Some are 550 metres tall. They are topped by monasteries, some of which are over 1,000 years old.

93

The Balkans

THE BALKANS ARE HOME TO VERY MANY DIFFERENT peoples, including Serbs, Croatians, Slovenians, Montenegrins, Romanians, Bulgars, Turks, Roma (Gypsies), Albanians and Greeks. The region has had a very troubled political history, largely as a result of the neglect and division it experienced during the hundreds of years it was ruled by the Austrian or Ottoman (Turkish) empires.

During the 1990s there was grim, often racist conflict in the northwest as a communist federation called Yugoslavia split up into separate countries known as Slovenia, Croatia, Bosnia-Herzegovina, Yugoslavia (Serbia-Montenegro) and Macedonia (which is also the name of the northernmost province of Greece).

In the northwest, the people of Romania overthrew a self-styled communist dictator, Nicolae Ceauşescu (1918–89), in 1989. Communist rule collapsed in Bulgaria in 1990 and in Albania in 1992. The region faces huge problems and great poverty. Greece too has faced many problems, including a period of right-wing military rule from 1967 until 1973, and disputes with its eastern neighbour, Turkey, and with the new Slavic republic of Macedonia to the north.

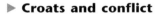

▲ **Defending Greece**
The ceremonial uniform of the Greek national guards, the Evzones, is made up of a tasselled cap, a kilt and white leggings. It is based on the dress of the mountain troops who fought against Turkey in the 1820s.

▲ **Olives**
Olives have been a symbol of Greek wealth since ancient times. Many are exported through the seaport of Kalamáta.

▶ **Croats and conflict**
An elderly Croatian man weaves canes to make a basket. Between 1991 and 1995 Croatia fought the Serbs in both Yugoslavia and Bosnia. Many Serbs living inside Croatia fled the country.

▼ **Ocean harvest**
Squid and octopus provide an important ingredient of many Greek meals and snacks. They are served at coastal restaurants and bars during the summer months.

FACTS

MACEDONIAN REPUBLIC
(Former Yugoslav Republic)
Republika Makedonija
Area: *25,713 sq km*
Population: *2.0 million*
Capital: *Skopje*
Other cities: *Bitola, Prilep*
Highest point: *Korabit (2,751 m)*
Official language: *Macedonian*
Currency: *Dinar*

ROMANIA
România
Area: *238,391 sq km*
Population: *22.6 million*
Capital: *Bucharest*
Other cities: *Brasov, Constanta*
Highest point: *Moldoveanu (2,543 m)*
Official language: *Romanian*
Currency: *Leu*

SLOVENIA
Republika Slovenija
Area: *20,256 sq km*
Population: *2.0 million*
Capital: *Ljubljana*
Other cities: *Maribor, Jesenice*
Highest point: *Triglav (2,863 m)*
Official language: *Slovenian*
Currency: *Tolar*

YUGOSLAVIA
(Serbia-Montenegro)
Savezna Republika Jugoslavija
Area: *102,173 sq km*
Population: *10.6 million*
Capital: *Belgrade*
Other cities: *Novi Sad, Nis*
Highest point: *Daravica (2,656 m)*
Official language: *Serbo-Croat*
Currency: *Dinar*

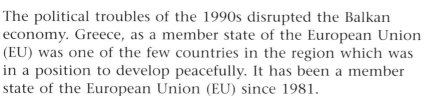

◄ ▲ Melon market
The pink, moist flesh of a Greek watermelon, or karpouzi, makes an ideal refreshment for a hot summer day.

The political troubles of the 1990s disrupted the Balkan economy. Greece, as a member state of the European Union (EU) was one of the few countries in the region which was in a position to develop peacefully. It has been a member state of the European Union (EU) since 1981.

The climate of the region is suitable for growing fruits, such as plums in the north and lemons and olives in the south, watermelons, maize, tobacco and sunflowers. Bulgaria grows roses for the perfume industry. The Balkan lands produce iron and steel, chemicals, machinery, textiles, clothing and footwear. Tourism is a major industry in Greece, which offers blue seas, island beaches and pretty, whitewashed villages. Recent political troubles disrupted what had also been a thriving tourist industry on the coast and islands of Croatia.

▲ Petals for sale
Attar of roses is an important product of Bulgaria's Kazanluk district. This is an oil made from rose petals. It is used to make perfumes.

▲ Threshing corn
Slovenian women thresh a harvest of corn. Many people still work on farms in rural areas of the Balkans. Hi-tech machinery is rare on these small, family-run farms.

▲ A volcanic island
The Greek island of Santorini, or Thíra, lies just to the north of Crete. Today, it attracts tourists, but about 3,500 years ago it was the scene of one of the biggest volcanic explosions in recorded history.

► The welder
Serbian metal industries are based in Pancevo, Kragujeva and Nis. The capital Belgrade is a centre of both heavy and light engineering. Output has continued despite the region's political strife and warfare in the 1990s.

The Balkans

THE EARLY HISTORY OF THE BALKANS IS MARKED BY the amazing civilizations that developed on the island of Crete from about 2500BC until 1400BC, and on the Greek mainland and islands from 1000BC onwards. The city states of ancient Greece produced great playwrights and poets, sculptors, athletes, soldiers, traders, mathematicians and thinkers. The idea of democracy – rule by a public assembly instead of by kings or dictators – was first tried out in Athens. Greek seafarers also colonized many other parts of the Mediterranean region.

In 338BC the northern kingdom of Macedonia conquered the rest of Greece, and a young ruler called Alexander the Great went on to found a vast empire which stretched southwards to Egypt and eastwards to India. This broke up after he died, and in 146BC Greece became part of the Roman empire, which soon swallowed up the whole Balkan region.

The eastern part of the Roman empire survived after the western part collapsed, and this developed into the Byzantine empire, with its capital at Constantinople (modern Istanbul, in the European part of Turkey). The Byzantine empire became a centre of Eastern Orthodox Christianity and of the Greek language and way of life.

▲ **The Parthenon**
Still crowning the Acropolis in Athens, the Parthenon was completed in 438BC. It was temple to Athena, the goddess of the city.

▲ **Bulgar tradition**
Traditional costume may still be worn for special festivals or dances in Bulgaria.

◀ **A great thinker**
Socrates was condemned to death in 399BC, for encouraging young people to ask awkward questions and disrespect the gods. He was one of the greatest philosophers of ancient Greece.

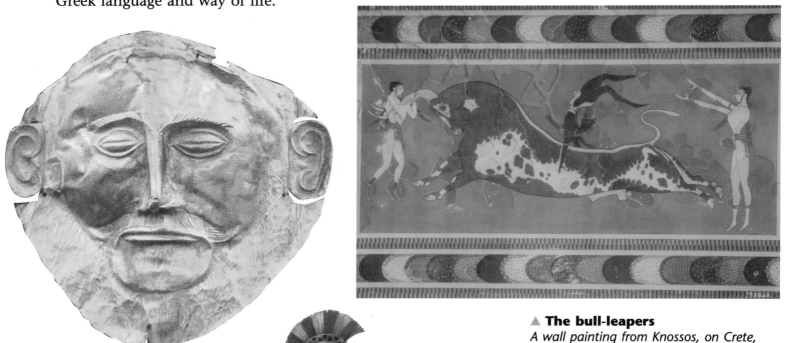

▲ **The bull-leapers**
A wall painting from Knossos, on Crete, shows acrobats vaulting over the backs of bulls. About 3,500 years ago, Knossos was the capital of the Minoan civilisation.

▲ **Mask of Mycenae**
This golden death mask comes from Mycenae, a citadel town in southern Greece. It was the centre of a civilization between 1900BC and 1100BC.

◀ **Greek armies**
Ancient Greece was never a united state until 338BC. When its armies did join forces, they defeated the mighty Persian empire. The Greek state of Sparta was particularly famous for its strict military training.

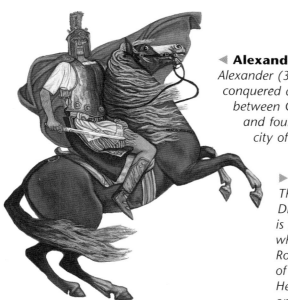

◀ Alexander the Great
Alexander (356–323BC) conquered all the lands between Greece and India, and founded the Egyptian city of Alexandria.

▶ Dracula's castle
The legend of Dracula, the vampire, is based upon Vlad IV, who ruled part of Romania in a reign of terror (AD1455–62). He impaled his victims on pointed stakes.

It fell to invading Turks in AD1453. Most of the Balkans remained under Turkish rule for over 400 years, with Austria-Hungary controlling parts of the northwest. World War I (1914–18) was started when a Serbian nationalist assassinated the heir to the Austrian throne in Sarajevo.

World War II (1939–45) brought Italian and German invasions of the Balkans and long years of guerrilla warfare, which continued after the war in battles between communists and monarchists. Communists failed to gain control of Greece, but succeeded elsewhere. However all the communist goverments in the Balkans collapsed in the 1980s and 90s, giving way to a new period of war and conflict in the north.

▲ A fateful day
On June 28, 1914 the Archduke Franz Ferdinand of Austria and his wife were assassinated in Sarajevo, Bosnia. This incident triggered World War I.

▲ Sophia cathedral
Bulgaria is part of the Eastern Orthodox Church. This domed cathedral shows the influence of the Byzantine empire.

◀ Slobodan Milosevic
Serbian nationalist Milosevic rose to power in the 1980s. In the 90s, he led campaigns against Bosnian Muslims and the ethnic Albanians of Kosovo.

AD	TIMELINE
330	Roman empire split: the eastern capital, Constantinople, goes on to become centre of Byzantine empire
500s	Slavs invade Balkans (through 600s)
680	Bulgars invade Bulgaria
1355	Death of Stephen Dushan, Serbia at height of power
1396	Turks invade Bulgaria
1453	Turks capture Constantinople, go on to invade western Balkans
1821	Greek War of Independence (until 1830)
1908	Bulgarian declaration of independence
1912	Balkan Wars (until 1913) drive Turks from most of Europe; Albanian independence
1914	World War I starts in Sarajevo: Turkey joins Central powers (Germany, Austria); Greeks join Allies 1917
1918	Formation of Yugoslavia
1939	World War II (until 1945): 1940, Italy invades Albania; Greece invaded by Axis troops (Italy, Germany, Bulgaria); Germany defeated
1946	Communist governments take power in Balkans; Greek communists defeated in civil war (until 1949)
1961	Albanian communist leader Enver Hoxha breaks with Soviet Union
1967	Military coup in Greece
1975	Greece becomes democratic republic
1981	Greece joins EEC (later EU)
1989	Violent revolution in Romania
1990	Collapse of communist governments in Balkans (through 1991), break-up of Yugoslavia
1990s	Balkan wars and ethnic persecution

The AMERICAS

North and South America are two great continents, lying between the Atlantic and Pacific Oceans. They are joined together by Central America, which narrows to a strip of land called the Isthmus of Panama. This is cut in two by the Panama Canal, which links the two oceans. Tropical Central America is bordered to the east by the Caribbean Sea, which is enclosed by the islands of the Greater and Lesser Antilles.

▲ **Cactus**

North America is dominated by a mountain system called the Rockies, which stretches from north to south. In the far north, frozen tundra yields to forest. To the south lie the prairies – rolling grasslands which are now used for grazing or crops, part of a great plain drained by the Mississippi-Missouri river system. In the southern United States and Central America are rocky canyons, wetlands, volcanoes and deserts.

In South America, the Rockies are matched by the mighty Andes mountain chain. To the east, a maze of waterways flows into the river Amazon, which is surrounded by the world's largest tropical rainforests. South American grasslands include the Llanos of Venezuela and the Argentinean Pampa. There are windy plateaus and coastal deserts. The continent ends in the bleak island of Tierra del Fuego, which reaches out like a claw towards Antarctica.

▲ **Boatbuilder,
Barbados**

▼ **Lake Moraine, Canada**

◀ **Sailfish**

◀ **Toltec
temple,
Mexico**

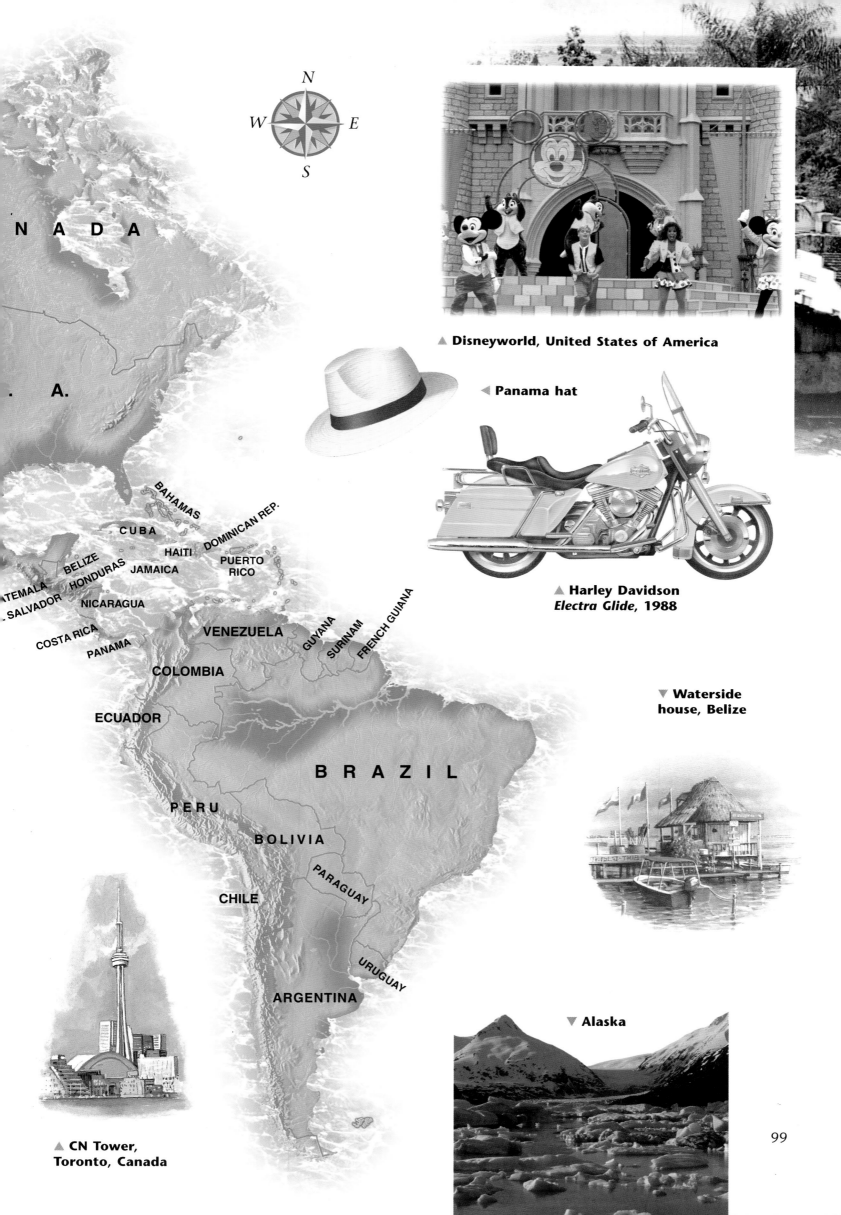

N A D A

. A.

BAHAMAS

CUBA

HAITI

DOMINICAN REP.

BELIZE

HONDURAS

JAMAICA

PUERTO RICO

TEMALA

SALVADOR

NICARAGUA

COSTA RICA

PANAMA

VENEZUELA

GUYANA

SURINAM

FRENCH GUIANA

COLOMBIA

ECUADOR

B R A Z I L

P E R U

BOLIVIA

PARAGUAY

CHILE

URUGUAY

ARGENTINA

▲ Disneyworld, United States of America

◄ Panama hat

▲ Harley Davidson
Electra Glide, 1988

▼ Waterside
house, Belize

▼ Alaska

▲ CN Tower,
Toronto, Canada

99

Canada and Greenland

Greenland is the world's biggest island. In fact, it is really several islands, which are welded together by a cap of permanent ice. This is up to 3,000 metres deep in places. Icebergs break off from its glaciers and float out to sea. Only one-sixth of Greenland is ice-free.

▼ **Tundra life**
The Arctic tundra supports a wide range of wildlife, including birds, such as the ptarmigan, and the caribou, a North American reindeer.

Across the Davis Strait lies Canada, by area the world's second largest nation. Its islands are scattered like pieces of a jigsaw puzzle across the Arctic, between the Beaufort and Labrador Seas. The tundra, a wide expanse of ice and snow, remains frozen all year round. However during the brief summer, the snow melts and lies in pools which attract insects and migrating birds.

Most of the Canadian mainland is taken up by a broad belt of forests and glacial lakes. A vast slab of ancient rock, the Canadian Shield, borders Hudson Bay. In the west are the high peaks of the Mackenzie, Rocky and Coast ranges, which descend to the moist, green coast of British Columbia.

The southern border with the United States crosses the prairies and the Great Lakes. The St Lawrence River and Seaway link Lake Ontario with the Atlantic Ocean. Here, the warm Gulf Stream meets the cold Labrador Current, bringing fog to the waters off Newfoundland, Canada's Maritime provinces and the little French islands of St Pierre and Miquelon.

▲ **In the Rockies**
Moraine Lake may be visited in the Banff National Park, in Alberta. Its vivid blue colour is caused by a silt known as 'rock flour.'

FACTS

CANADA
Area: 9,970,610 sq km
Population: 30.0 million
Capital: Ottawa
Other cities: Toronto, Montréal, Vancouver
Highest point: Mount Logan (5,951 m)
Official languages: English, French
Currency: Canadian dollar

GREENLAND
Kalaallit Nunaat – Grønland
SELF-GOVERNING ISLAND TERRITORY OF DENMARK
Area: 2,175,600 sq km
Population: 0.06 million
Capital: Nuuk (Godthåb)
Highest point: Gunnbjørn Fjeld (3,700 m)
Official languages: Danish, Greenlandic (Kalaallisut)
Currency: Danish krone

ST PIERRE AND MIQUELON
TERRITORIAL COLLECTIVITY OF FRANCE
Area: 241 sq km
Population: 6,500
Capital: Saint-Pierre
Official language: French
Currency: French franc

▶ The beaver

Beavers live on many Canadian lakes. They use their powerful gnawing teeth to fell trees and shred wood.

▶ Tundra soil

A pingo is a mound of soil filled with a core of expanding ice. When the ice melts, the pingo collapses.

GREENLAND

N
W E
S

CANADA

ARCTIC OCEAN

LINCOLN SEA

Ellesmere Island

Denmark Strait

GREENLAND

Melville Island

Devon Island

BAFFIN BAY

Banks Island

Prince of Wales Island

BEAUFORT SEA

Victoria Island

Baffin Island

Davis Strait

LABRADOR SEA

ALASKA (U.S.A.)

Dawson

Norman Wells

Great Bear Lake

FOXE BASIN

MACKENZIE MOUNTAINS

Mackenzie

YUKON TERRITORY

▲ Mt. Logan 5,951 m

Whitehorse

NORTHWEST TERRITORIES

Southampton Island

Hudson Strait

HORN MOUNTAINS

Yellowknife

Dubawnt Lake

Coats Island

Mansel Island

Ungava Peninsula

Liard

Great Slave Lake
Fort Resolution

HUDSON BAY

Fort Smith

ROCKY MOUNTAINS

CARIBOU MOUNTAINS

Lake Athabasca

Churchill

La Grande Rivière

NEWFOUNDLAND

Goose Bay

BRITISH COLUMBIA

Peace

CANADA

Reindeer Lake

Churchill

Belcher Islands

Feuilles

ince Rupert

Prince George

Peace River

Nelson

JAMES BAY

OTISH MOUNTAINS

Gander

EEN LOTTE NDS

ALBERTA

MANITOBA

Akimiski Island

Newfoundland St John's

Edmonton

Severn

Peribonca

Anticosti Island

Fraser

Red Deer

N. Saskatchewan

Prince Albert

Lake Winnipeg

QUEBEC

St. Lawrence

Gulf of St. Lawrence

Vancouver Island

Kamloops

Saskatoon

Lake Winnipegosis

Albany

ONTARIO

PRINCE EDWARD ISLAND

Calgary

SASKATCHEWAN

Quebec

NEW BRUNSWICK

Charlottetown

Vancouver

Medicine Hat

Lake Manitoba

Lake Nipigon

Montreal

Fredericton

St John

NOVA SCOTIA

Victoria

S. Saskatchewan

Regina

Winnipeg

Thunder Bay

Ottawa

Halifax

UNITED STATES OF AMERICA

Lake Superior

Georgian Bay

Lake Huron

Toronto

Lake Ontario

Niagara Falls

ATLANTIC OCEAN

Hamilton

Windsor

Lake Erie

◀ Takkakaw Falls

Takkakaw means 'wonderful' in the language of the Cree people. The falls tumble over a drop of 254 metres in the Yoho National Park, British Columbia.

▶ Bath time!

The moose can stand 2.3 metres at the shoulder. This one is pictured at Pukaskawa National Park, Ontario, on the north shore of Lake Superior.

◀ Musk ox

101

Canada and Greenland

GREENLAND IS A DEPENDENCY OF DENMARK, but has been self-governing since 1981. With the exception of Antarctica, it is the world's least-populated land. Temperatures on the central ice cap can drop below –65°C, but ocean currents keep the southwest coast relatively mild. The fishing industry is a major employer.

In Canada, too, the severe climate of the north has restricted settlement. Seventy-seven percent of Canadians are town-dwellers and the big cities are all in the south, where the climate is milder and transport easier. The Canadian capital is Ottawa, a city in southeastern Ontario. The commercial centres are the much larger cities of Toronto and Montréal.

Canada is an independent nation whose head of state is the British monarch. It is organized on federal lines, with provinces and territories. Canada was a founder member in 1994 of the North American Free Trade Agreement (NAFTA), which strengthened economic ties with the United States and Mexico to the south.

▲ Sky level
Toronto's CN Tower, over 553 metres high, is the world's tallest structure. Its viewing platform is a dizzying 440 metres above the ground.

▼ Lumberjack years
The loggers who felled the Canadian forests a century ago had few mechanical aids. They lived hard lives in remote camps.

▲ In search of cod
The Grand Banks are shallow waters off the coast of Newfoundland. They have attracted international fishing fleets for about five centuries, but stocks of cod have declined in recent years.

◄ Fighting the blaze
A helicopter moves in to fight a forest fire. Canada's vast conifer forests stretch in a broad belt across central and northern regions, from British Columbia to the Laurentian Plateau.

► Logging today
Canada has over 453 million hectares of forest and is the world's leading exporter of forest products. These include timber in various forms, wood pulp and paper.

◀ Toronto skyline
With a population of
nearly five million, Toronto
is Canada's biggest city.
It is a national centre
of business and
communications.

◀ Space tech
Canada has an
important aerospace
industry. It built this
robotic lifting device,
called the Canadarm,
for use on American
Space Shuttles.

▼ Oil rig, Alberta
Over 90 percent of
Canada's reserves of
oil and natural gas
are found in Alberta
province. Major
production regions are
at Lloydminster, Fort
McMurray and here,
at Cold Lake.

The Canadian prairies supply wheat to the world
and provide pasture for cattle. The great forests
send timber to the sawmills. There are reserves of
oil, natural gas, copper, gold, iron ore and nickel,
and there are plenty of rivers and lakes to provide
hydroelectric power. Factories manufacture cars,
paper, steel and chemicals.

Food products include maple syrup, apples, cheese
and beer. Until recently Newfoundland lay off one
of the richest fishing grounds in the world, but
overfishing has led to dwindling stocks and a
ban on trawling until numbers recover.

Like the United States, Canada is a melting
pot of peoples and cultures. For the last
30 years Canadian politics have been
dominated by the future of Québec
province, where a large number of
French-speakers wish to break
away from the rest of Canada
altogether. English-speaking
Canadians share many of the
interests of their American
neighbours – but are always
keen to emphasise their own
independent way of life.

▲ ▶ Sugar maple
Maple syrup is boiled up from
a sweet sap, collected from the
maple tree. It was invented long
ago, by native peoples of the St
Lawrence valley. Today it is
popular in Canada and the
United States, where it is poured
over pancakes and waffles.

▲ In Montréal
Montréal, capital of Québec
province, is the chief city of
French-speaking Canada and a
centre of commerce and the arts.
Shop signs are often in French.

Canada and Greenland

THE FIRST PEOPLE TO SETTLE Canada were prehistoric peoples from the Asian region of Siberia. Their descendants include the First Peoples who live in Canada today – groups such as the Mohawk, Micmac, Innu, Cree, Dene and Kwakiutl.

▲ Across the Arctic
Vehicles fitted with skis, such as skidoos, have become the most popular method of travel for the Inuit and other peoples of the north.

They were followed by waves of Inuit hunters, who set up scattered settlements across the Canadian Arctic and Greenland. Except in Greenland, the descendants of all these native peoples are today greatly outnumbered by later immigrants. They have faced a long struggle to gain rights to their own land, but there have been successes, too. A huge but very sparsely-populated area of Canada became the Inuit territory of Nunavut in 1999.

▲ Erik Cove, Québec, 1904
A hundred years ago, hunting met all the Inuits' needs, providing meat, hide and fur for clothing, bones for needles and tools and gut for thread.

Vikings from Scandinavia arrived in Canada and Greenland about 1,000 years ago. They soon lost their foothold in Canada, but stayed in Greenland for about 500 years. Danish traders returned to Greenland in the 1700s and it later became a Danish colony.

◄ Canadian Pacific
This railway was Canada's first continental crossing. It was completed in 1885 and ran between Montréal and Port Moody, Vancouver.

▼ Chuck wagon race
These are a feature of North America's toughest rodeo, the Calgary Stampede. Held each year in mid-July, this celebration of cowboy skills dates back to 1912.

► Totem poles
Tall poles carved from cedarwood were erected in villages of Canada's Pacific coast by native chiefs. They represented guardian spirits and histories of the family or tribe.

▼ Monster mash
In a different kind of stampede, heavy-metal racers with big tyres show off in Calgary, Alberta.

▼ The champions
Ice hockey was first played in Canada in the 1850s and today is the country's most popular sport.

▶ Winterlude
The Winterlude festival is held each February in Ottawa. Events include sculpture in ice and snow, and skating on the Rideau Canal.

It was the French who explored and colonised the lands around Canada's St Lawrence River in the 1500s, while the British went on to trade around Hudson Bay. In the 1700s the two nations fought bitterly for control of Canada. The British won. Their numbers grew after the 1770s, as families loyal to Britain fled northwards from the newly independent United States.

The French retained their language and their Roman Catholic faith, but by 1867 Canada was united as a dominion within the British empire. Many more peoples settled in Canada as the nation expanded across the prairies towards British Columbia. Newfoundland and Labrador joined Canada in 1948.

Although people of British (especially Scots) and French descent still make up a large percentage of the Canadian population, it has grown in the last two centuries to include Ukrainians, Dutch, Russians, Poles, Germans, Italians, Chinese, Indians, Vietnamese and Afro-Caribbeans.

◀ Newfoundlanders
Fishing is a way of life on the remote island of Newfoundland, off the Atlantic coast. The islanders voted to join Canada in 1949.

▶ Dogsled, Québec
Teams of dogs pulled sleds in the old Arctic and are still used today. Dogsled racing is a popular sport.

◀ The 'Mounties'
The Northwest Mounted Police, with their red coats and distinctive hats, were founded in 1873. They tamed the Wild West. In 1920 they became a national force, the Royal Canadian Mounted Police.

AD	TIMELINE
900s	Vikings discover and settle Greenland
1410	Greenland colonists lose contact with Scandinavia
1497	John Cabot discovers Newfoundland
1534	Jacques Cartier explores St Lawrence River
1608	Samuel de Champlain founds Québec
1642	Montréal founded
1670	Hudson's Bay Company established
1713	British gain Newfoundland
1759	British defeat French at Québec
1763	Canada becomes a British colony
1776	Founding of KGH, the Royal Greenland Trading Company
1814	Greenland becomes Danish colony
1840	Act of Union joins Upper and Lower Canada
1867	Dominion of Canada: Ontario, Québec, Nova Scotia, New Brunswick
1870s	Manitoba, British Columbia, Prince Edward Island join Canada
1896	Klondike gold rush (until 1898)
1905	Alberta, Saskatchewan join Canada
1914	World War I (until 1918): Canada joins Allies
1939	World War II (until 1945): Canada joins Allies
1949	Newfoundland joins Canada
1959	St Lawrence Seaway opened
1968	Separatists demand free Québec
1979	Home rule for Greenland
1994	Canada in North American Free Trade Agreement (NAFTA)
1995	Québec referendum rejects separatism
1999	Self-governing homeland for Inuit (Nunavut)

United States of America

▲ **Maize**

Central North America is dominated by the United States of America (USA). The only other territory is a group of 150 small islands, lying 1,120 kilometres to the southeast of the state of New York, in the North Atlantic Ocean. These make up a British colony called Bermuda.

▲ **Niagara Falls**
These spectacular waterfalls are on the United States-Canada border. Not only a tourist attraction, they are also a major source of hydroelectric power.

Forty-eight of the United States lie between the Canadian and Mexican borders. This is the American heartland, with coastlines on both the North Atlantic and North Pacific oceans. It is a land of amazing variety, where long, lonely highways cut through deserts and mountains, farmland and forest, linking together the big cities of the east and west.

To the north, beyond Canadian territory, lies Alaska. This includes great areas of Arctic wilderness, forming the largest state of all. Its islands are inhabited by large grizzly bears and its waters by schools of whales.

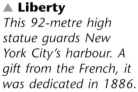

▲ **Liberty**
This 92-metre high statue guards New York City's harbour. A gift from the French, it was dedicated in 1886.

Far to the west, the volcanic Hawaiian islands form a Pacific outpost of the USA. Tourists come here to enjoy the warm climate and to see the island's spectacular volcanoes.

▲ **Bald eagle**

The area of the United States as a whole makes it the fourth largest country in the world, after the Russian Federation, Canada and China.

◀ **Bryce Canyon**
Pink, orange and buff-coloured rocks have been eroded by wind, water and ice into extraordinary pinnacles at this national park, in Utah.

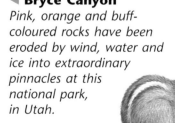

▲ **Chipmunk**

FACTS

UNITED STATES OF AMERICA
Area: 9,363,520 sq km
Population: 265.3 million
Capital: Washington DC
Other cities: New York City, Los Angeles, Chicago
Highest point: Mt McKinley (6,194 m)
Official language: English
Currency: US dollar

BERMUDA
British Overseas Territory
Area: 53 sq km
Population: 0.06 million
Capital: Hamilton
Official language: English
Currency: Bermuda dollar

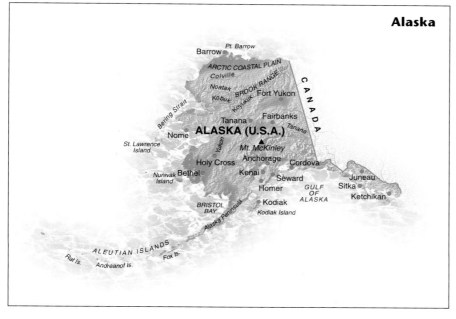

Alaska

Pt. Barrow
Barrow
ARCTIC COASTAL PLAIN
Colville
Noatak BROOK RANGE Fort Yukon
Kobuk Koyukuk
Bering Strait Tanana Fairbanks
Nome **ALASKA (U.S.A.)** Tanana
St. Lawrence Yukon
Island Mt. McKinley
Holy Cross Anchorage
Nunivak Bethel Kenai Cordova
Island Seward GULF
Homer OF Juneau
ALASKA Sitka
Kodiak Ketchikan
BRISTOL
BAY Kodiak Island
Alaska Peninsula
ALEUTIAN ISLANDS
Rat Is. Fox Is.
Andreanof Is.

▲ **The largest state**
Alaska is a land of misty shores, islands, towering mountains, glaciers and remote, snowy wilderness.

UNITED STATES OF AMERICA

N
W E
S

CANADA

Seattle Spokane Kalispell Havre
Olympia WASHINGTON Missouri Grand Falls Williston Minot Grand Forks MINNESOTA Lake Superior MAINE
Mt. Rainier Missoula Helena NORTH Jamestown Duluth Marquette Bangor
4,392 m Pendleton Butte MONTANA Bismarck DAKOTA Fargo St. Cloud WISCONSIN Lake Huron Augusta
land Lewiston Billings Aberdeen Minneapolis Green Bay MICHIGAN Burlington Portland
OREGON IDAHO Sheridan St Paul La Crosse Grand VERMONT NEW
Klamath Falls Boise Idaho Black James Sioux Milwaukee Rapids Montpelier HAMPSHIRE Concord
Falls Hills SOUTH Falls Madison Lansing Rochester Syracuse Boston MASSACHUSETTS
Twin Falls Snake WYOMING Rapid City DAKOTA Mississippi Detroit Buffalo Albany Hartford Providence
Redding Pocatello Casper Pierre IOWA Chicago Rockford Windsor Cleveland Erie Scranton RHODE ISLAND
Reno Elko Great Ogden Rock Springs Laramie Cedar Rapids Gary Toledo OHIO Akron PENNSYLVANIA Philadelphia New York City
Salt Lake Scottsbluff Norfolk Sioux City Davenport Columbus Pittsburgh Harrisburg Baltimore NEW JERSEY
Carson City Salt Lake City Fort Collins NEBRASKA Omaha Des Moines Peoria INDIANA Dayton WEST WASHINGTON D.C. Annapolis Dover
NEVADA UTAH Cheyenne Grand Island ILLINOIS Indianapolis Cincinnati VIRGINIA Chesapeake Bay DELAWARE
GREAT BASIN Boulder Lincoln Springfield Louisville Frankfort Charleston VIRGINIA Richmond MARYLAND
Mt. Whitney Denver S. Platte Kansas Jefferson St. Louis Lexington Roanoke Norfolk
4,418 m Provo COLORADO Colorado City City KENTUCKY Greensboro
CALIFORNIA Death Grand Mt. Elbert Springs Abilene Topeka MISSOURI Evansville Knoxville Winston- Raleigh Cape Hatteras
Valley Junction 4,399m Salina Paducah Salem NORTH CAROLINA
Bakersfield Las Vegas Grand Monument Pueblo Arkansas Hutchinson Springfield Nashville TENNESSEE Charlotte
Canyon Valley Durango Wichita Joplin Chattanooga Greenville
os Angeles San Bernardino Flagstaff Trinidad KANSAS Memphis SOUTH CAROLINA Cape Fear
ISLANDS NEW MEXICO Cimarron OKLAHOMA ARKANSAS Tennessee Alabama Greenville Columbia
Long Beach ARIZONA Gallup Santa Fe Canadian Tulsa Fort Smith Birmingham APPALACHIAN Augusta Charleston
San Diego Phoenix Albuquerque Oklahoma Arkansas Little Rock Tupelo ALABAMA Atlanta Macon
Yuma Mesa Amarillo City Red River Shreveport Meridian Columbus GEORGIA Savannah
Tucson Lubbock LOUISIANA Montgomery Albany
MEXICO Las Cruces Midland Wichita Falls Texarkana Jackson MISSISSIPPI Mobile Jacksonville
Douglas El Paso Odessa Abilene Dallas Brazos Fort Worth Alexandria Baton Rouge Biloxi Tallahassee St. Augustine
San Angelo TEXAS Waco Beaumont New Orleans Pensacola Daytona Beach
Pecos Austin Houston Port Arthur Mississippi Delta Tampa Orlando Cape Canaveral
Rio Grande San Antonio Galveston St. Petersburg FLORIDA West Palm Beach
Corpus Christi GULF OF MEXICO Fort Myers Lake Miami
Laredo Okeechobee
Brownsville Key West Florida Keys
Straits of Florida

▼ **Organ Pipe Cactus National Monument**
Summer temperatures can soar to 41°C in this part of Arizona. Spiny cacti, bearing pink flowers in summer, line the Ajo Mountain drive.

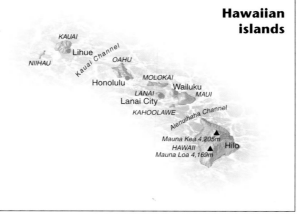

Hawaiian islands

KAUAI
Lihue
NIIHAU Kauai Channel OAHU
MOLOKAI
Honolulu LANAI Wailuku
Lanai City MAUI
KAHOOLAWE Alenuihaha Channel
Mauna Kea 4,205m
HAWAII
Mauna Loa 4,169m Hilo

BERMUDA

▲ **Sea cow**
The rare manatee is found along the Florida coast.

107

United States of America

THE NORTHEASTERN UNITED STATES IS A temperate land of green river valleys, woodland and rocky shores. The coastal plain is narrow in the north but widens to the south of Chesapeake Bay, where it is fringed by long, sandy beaches. The flat lands rise to the Appalachian system, which includes the Green, White, Allegheny and Blue mountain ranges. In the far southeast, the sunny peninsula of Florida ends in a string of small sandy islands, known as keys. Here, the calm can be shattered by savage hurricanes.

To the south of the Great Lakes lies a broad plain, drained by the Missouri and Mississippi rivers. The latter forms a muddy delta where it enters the steamy Gulf of Mexico. Flat, dusty farmland stretches westwards, bedevilled by whirlwinds called tornadoes. The American prairies, once a great sea of grass stretching to the badlands at the foot of the Rocky Mountains, are now patchworked with fields of crops, farm buildings and ranches.

▲ The Everglades
The Everglades is a vast system of wetlands which drains southern Florida. It is home to alligators, tree snails and egrets.

▼ Okefenokee swamp
This wildlife refuge lies on the Georgia-Florida border. It is a maze of waterways, marshes and floating islands. Its bald cypress trees are draped with Spanish moss.

▲ The colours of fall
New England – the far northeastern region of the United States – is famous for the beautiful reds, oranges and browns of its temperate woodlands, before the leaves fall in autumn.

▲ The Colorado Rockies
The Rocky Mountains National Park is entered from the town of Estes Park, Colorado, on the Big Thompson River. The massive ranges of the Rockies form the backbone of the North American continent.

◄ Golden Gate Bridge
This is California's most famous landmark. Opened in 1937, the bridge crosses a strait which connects San Francisco Bay with the open Pacific Ocean.

► Sea lions
Graceful in the water, comical on land, sea lions live on many islands off the Californian coast. Males can be up to three metres long.

▲ Fire warning
A notice in Bryce Canyon, Utah, warns of the danger of forest fires. America's vast forests are often devastated by terrifying blazes.

▼ Grand Canyon
In Arizona, the Colorado River has carved out the world's biggest gorge. In parts, it plunges to a depth of 1.5 kilometres.

Beyond the massive peaks of the Rockies lie dazzling salt flats and rocks worn into bizarre shapes by the wind and weather. It was the Colorado River which cut out the world's most impressive gorge, the Grand Canyon. The deserts of the southwest shimmer in the heat.

▲ Night-raider
The raccoon, with its banded eyes and tail, often raids dustbins by night.

The Coast, Cascade and Sierra Nevada ranges of the far west act as a barrier to rain-bearing winds from the ocean. However the western slopes catch the rain, making the forests of the northwest cool, moist and green. The north of California has a pleasant, spring-like climate when not shrouded in fog. The south is hotter and drier.

Alaska stretches high into the deep-frozen Arctic and includes empty wilderness, icy shores and some of the highest peaks in the Americas. The Hawaiian islands are really a chain of very high submarine volcanoes, with their tops emerging from the waves.

◄ Cliff cascade
The cliffs of California's Yosemite Valley are braided with beautiful waterfalls. This is the upper part of the Yosemite Falls, the world's second-highest, with a drop of 739 metres.

► Tornado twisters
Raging whirlwinds called tornadoes or twisters are common in the Prairie states. They can spin at 500 kilometres per hour.

▲ Lava flow
This slow-moving lava from a Hawaiian volcano is known as pahoehoe.

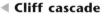

United States of America

DURING THE LAST 75 YEARS THE UNITED STATES HAS become the world's most powerful nation, and one of the richest.

It is a democratic republic which works on a federal system, with its states having powers to make their own laws. The head of state is a president, who is elected for a term of four years.

In addition to its 50 states, the USA also governs various island territories in the Pacific Ocean and the Caribbean Sea.

The United States is one of the world's great food producers, growing wheat, soya beans, maize, citrus fruits and vegetables. Coasts and rivers produce large catches of fish. Milwaukee is famous for its beers, Kentucky for its bourbon and California for its wines.

▲ **Mount Rushmore**
Completed in 1941, these giant heads show former presidents Washington, Jefferson, Roosevelt and Lincoln.

The United States has great mineral wealth. It drills for oil and natural gas in Texas, the Gulf of Mexico and Alaska. It is the world's largest producer of coal. It has gold, uranium, copper and iron ore. Dammed rivers and lakes are used to produce hydro-electric power. Managed forests provide timber, with the far northwest a major centre of logging.

▲ **All-American food**
What's on the menu? As well as fast food such as hamburgers and hot dogs, the United States has given the world cola drinks, breakfast cereals and Florida orange juice.

◄ **Early skyscrapers**
The world's first high-rise buildings were built in Chicago and New York City. Better known as the 'Flat Iron', New York's 20-storey Fuller Building was erected in 1901.

▲ **Texas oil**
Texas built its wealth on cattle, cotton and oil. Today, the United States ranks as the world's second largest oil producer.

▼ **Biosphere 2**
This experimental base was set up in the Arizona Desert. Inside, all of the Earth's habitats are recreated in miniature. It is a practice run for future space bases on other planets or on the Moon.

▲ **The 'Big Apple'**
New York City is home to over 7.5 million people. It is an exciting, high-speed town, a centre for finance, business and the arts.

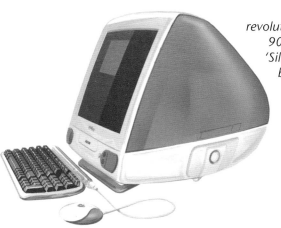

◄ Cyber success
The world computer revolution of the 1980s and 90s was spearheaded in 'Silicon Valley,' California by firms such as Apple and Microsoft.

► Disneyworld
This famous Florida theme park, opened in 1971, was the second one to be dreamt up by animated film pioneer Walt Disney.

American factories make a wide range of household goods, textiles and garments. California is the world centre of computer technology and software, and also the centre of the film and television industry. The United States is the leading manufacturer of aeroplanes and spacecraft. The city of Detroit makes cars.

◄ A Hollywood star
Stars are inset in the pavement in Hollywood, California, as a tribute to the big names of cinema. Walt Disney (1901–66) is honoured as the creator of Mickey Mouse.

Services such as banking, finance and insurance are now more important to the American economy than manufacturing. New York City's Wall Street is the centre of the American financial world. The United States, with neighbouring Canada and Mexico, set up the North American Free Trade Agreement (NAFTA) in 1994.

Although the United States has such great economic power, it too has its problems. It faces increasing competition from other countries and has large debts.

▼ Shop till you drop!
American ways of buying and selling have spread around the world. Shopping malls are an American invention.

▲ Kennedy Space Center
A space shuttle blasts off on another mission from this launch site on Cape Canaveral, Florida.

▲ Harley Davidson
Electra Glide, 1988

United States of America

THREE-QUARTERS OF ALL AMERICANS live in towns and cities.

▲ Skating in Central Park
A haven for relaxation, Central Park lies at the heart of New York City's busy Manhattan district. It is a place to walk, jog, rollerblade or picnic.

Despite the wealth of the nation as a whole, many people in both city and country areas suffer from poverty.

The USA has no official religion, and yet religion plays a very important part in everyday life. Eighty-four percent of the population is Christian, with Protestants outnumbering Roman Catholics two to one. Judaism is the faith of two percent.

English is the official language, with Spanish widely spoken in some areas. The nation's first peoples (Native Americans, Inuit and Aleuts) now make up only one percent of the population. However there has been a strong revival of interest in their traditional cultures and in their rights as citizens.

▲ A free press
The United States has many famous city-based newspapers, including the Washington Post *and the* New York Times.

▼ In Chinatown
Since the 1800s, the United States has had a large population of Chinese descent. In the Chinatown district of New York's Lower East Side, signs are in Chinese and shops sell Chinese food.

▲ Native American life
The first Americans still face many social and economic problems, but they have retained a fierce pride in their traditions and customs.

▲ Amish farmer

▼ Riding School, New Mexico
Horse-riding has been part of the American way of life since the days of the Wild West.

African Americans
America's large black population is of African descent. Afro-Americans have played an important part in the advancement of civil rights and in American culture. They invented many popular musical forms, including jazz and the blues.

Fourth of July
The United States marks its independence, won from Great Britain in 1776, each July 4th. This parade is taking place in the New England state of Maine.

The population as a whole comes from many different roots. The majority are Whites of European descent. These include English, Scots, Welsh, Irish, French, Italians, Germans, Dutch, Swedes and Poles. About nine percent are Hispanics, Spanish-speaking people originally from Mexico, Central America or the Caribbean. There are Jews, Chinese, Japanese, Koreans and Vietnamese. There are Polynesians from Hawaii and Samoa. About 12 percent of Americans are blacks of African and Caribbean descent, whose ancestors were enslaved by the European settlers during the first 300 years of settlement.

The minority peoples of the United States, from the Native Americans to the Afro-Americans, have all experienced racism and poverty. And yet the real wealth of the nation lies in the rich mixture of their cultures, in their music and dance, writing and art, customs and beliefs, food and festivals.

American writers, artists, musicians and film directors have had a great influence on the world during the last 100 years. Popular sports include baseball, basketball and American football.

American fast foods

Hawaiian dance
The Hawaiian islands became a state of the Union in 1959. The islanders' dances and traditional costume belong to the Polynesian culture of the Pacific.

Hot gospel
Christian Churches from the African tradition developed the choral sound known as gospel music, which also influenced popular music. This singer is also a New York City policewoman.

Home run
Baseball is one of America's most popular sports. It has been played since the 1840s.

American football
Fast and exciting, American football is popular both at college and at professional level. Superbowl contests have been known to attract world television audiences of over 138 million.

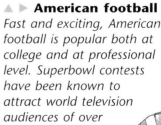
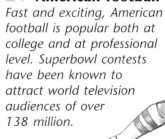

United States of America

THE NATIVE AMERICANS ARE DESCENDED from prehistoric peoples who crossed into North America from the Asian region of Siberia sometime after 30000BC. They developed many different cultures in different parts of the Americas.

In the 1500s colonists from Europe – English, Dutch, Spanish and French – began to explore and settle the east and south. At first they made alliances with the Native American peoples (whom they called 'Indians') but were soon entangled in bitter wars with them and with each other. Many Europeans purchased African slaves to work the land they had seized, planting tobacco and cotton.

▲ **Fort Pitt**
Nineteenth-century traders meet Native Americans at Fort Pitt. The Native Americans lost their lands and their way of life as European settlers moved West.

▲ **Warring states**
Flags of the Confederacy and the Union, the two conflicting sides in the American Civil War (1861–65).

◀ **Pilgrim Fathers**
The most famous European settlers were a group of Puritans, religious refugees from England, who founded a colony at Plymouth Rock, Massachusetts, in 1620.

The British colonists revolted against rule from London, declaring their independence in 1776. They defeated the British troops and founded a new republic. This survived a bitter civil war between the northern and southern states, which lasted from 1861–65 and brought an end to slavery.

▼ **A bitter conflict**
Over 360,000 Union (northern) troops were killed during the Civil War, and 260,000 Confederate (southern) troops. Thousands of civilians also died.

▲ **President Abe**
Abraham Lincoln was 16th President of the United States. He was assassinated after the Confederate defeat, in 1865.

▼ **Westward by wagon**
In the 1840s thousands of pioneers headed west along routes such as the Oregon Trail, which stretched for 3,200 km.

The rush for gold
In 1848 gold was discovered on the American River, in California. Soon prospectors were arriving from all over the world in the hope of making their fortune.

America mourns
John F Kennedy, a young and dynamic president, was assassinated in 1963. Here he is mourned by his wife Jackie and his brother, Robert.

The new country grew and grew. Pioneers pushed west into 'Indian Territory', seizing land and gunning down the Native American tribes. Vast areas of land were purchased from the French, Spanish and Russians (who governed Alaska until 1867). The east became industrialized, railways crossed the continent, gold was found in California. Thousands of poor people left Europe in search of fame and fortune in this new land.

The United States fought alongside the Allies in World War I, from 1917–18. After the boom years of the 1920s came an economic crash in 1929 and years of hardship. The United States was brought into World War II in 1941, when Japan bombed a naval base at Pearl Harbor, in the Hawaiian Islands.

Martin Luther King
This great campaigner for civil rights won the Nobel Peace Prize in 1964, and was murdered in 1968.

After the end of the war in 1945, relations with one former ally, the communist Soviet Union, went from bad to worse. During this 'Cold War', the two rival powers remained on the brink of hostilities until the late 1980s.

Political parties
The two most powerful political parties are the Democrats and Republicans. Here, Democrats hold their convention in Chicago.

AD	TIMELINE
1565	Spanish found St Augustine, Florida
1607	English found Jamestown, Virginia
1620	Pilgrim Fathers found Plymouth colony
1700s	Height of slave trade
1773	'Boston Tea Party'
1775	American Revolution
1776	Declaration of Independence
1783	Treaty of Paris: Britain loses colonies
1789	George Washington first president
1791	Bill of Rights
1803	Louisiana Purchase
1812	War with Britain (until 1814)
1819	Florida bought from Spain
1846	War with Mexico (until 1848): California and New Mexico gained
1861	Civil War (until 1865)
1865	President Lincoln assassinated
1867	Alaska bought from Russia
1876	Battle of the Little Big Horn
1890	Massacre at Wounded Knee
1898	Spanish-American War
1917	USA joins Allies against Central Powers in World War I (until 1918)
1929	Wall Street Crash: economic crisis
1941	Japan attacks Pearl Harbor: USA enters World War II on side of Allies (until 1945)
1950s	'Cold War' hostility towards Soviet Union (until 1990)
1961	USA in Vietnam War (until 1973)
1963	President John F Kennedy assassinated
1994	USA enters North American Free Trade Agreement (NAFTA)
1992	Bill Clinton elected President (re-elected 1996)

Mexico and Central America

▶ Hummingbird
Feathers shimmer as this tiny bird sips nectar from a tropical flower.

Mexico's northern border runs along the banks of a river known in the United States as the Rio Grande and in Mexico as the Rio Bravo del Norte. It passes through hot, dusty desert.

A long, thin peninsula, Baja (Lower) California, extends southwards from Tijuana into the warm, blue Pacific Ocean. The Western and Eastern Sierra Madre ranges enclose a high central plateau, which includes deserts, lakes, swamps and smoking volcanoes. Violent earthquakes are common. Southern ranges include the Southern Sierra Madre and the Chiapas Highlands. In the far southeast, the Yucatán Peninsula forms a broad hook around the Bay of Campeche. In the south of the country, the vegetation includes lush, tropical forest.

The seven small countries of Central America lie to the south of the Isthmus of Tehuantapec. The landmass snakes from northwest to southeast, reaching its narrowest point at the Isthmus of Panama. Its backbone is a series of peaks, including many active volcanoes, and high plateaus or mesas. This highland chain is broken only by the expanse of Lake Nicaragua, which covers an area of 8,430 square kilometres. The Caribbean coastal strip is low-lying and flat, with swamps and lagoons. The Central American climate is tropical and often hot and humid.

◀ Saguaro cactus
The saguaro of Mexico's north-western deserts is the world's biggest cactus, growing to 15 metres or more.

▲ Panama Canal
This vital shipping link was opened in 1914. It cuts through the Isthmus of Panama to link the Atlantic and Pacific.

◀ Under the volcano
Costa Rica is a mountainous country. The Arenal Volcano, in the Cordillera de Guanacaste, last erupted in 1968.

FACTS

BELIZE
Republic of Belize
Area: 22,696 sq km
Population: 0.2 million
Capital: Belmopan
Other cities: Belize City, Dangriga
Highest point: Victoria Peak (1,122 m)
Official language: English
Currency: Belize dollar

COSTA RICA
República de Costa Rica
Area: 51,100 sq km
Population: 3.4 million
Capital: San José
Other cities: Limón, Alajuela
Highest point: Chirripo Grande (3,819m)
Official language: Spanish
Currency: Costa Rican colon

EL SALVADOR
República de El Salvador
Area: 21,041 sq km
Population: 5.8 million
Capital: San Salvador
Other cities: Santa Ana, San Miguel
Highest point: Monte Cristo (2,418 m)
Official language: Spanish
Currency: Salvadorean colon

GUATEMALA
República de Guatemala
Area: 108,889 sq km
Population: 10.9 million
Capital: Guatemala City
Other cities: Puerto Barrios, Quezaltenango
Highest point: Tajumulco (4,220 m)
Official language: Spanish
Currency: quetzal

▶ **In Belize**
The low-lying, swampy coast of Belize is fringed by coastal shallows and islands called cays. Hurricanes are common in August and September.

▲ **Chichén itzá**
This ruined city in Mexico's Yucatán peninsula was a centre of the Mayan and Toltec civilizations between AD800 and 1180.

MEXICO

BELIZE

▲ **After the hurricane**
Much of Central America, including the Honduran capital, Tegucigalpa, was devastated by Hurricane Mitch in 1998.

GUATEMALA   HONDURAS

▲ **Howler monkey**
Some species of these large, noisy monkeys are threatened by the loss of tropical forests in Central American rainforests.

◀ **Toucan**
This bird's huge bill is used to eat tropical fruits.

▲ **Monarch butterfly**

EL SALVADOR

COSTA RICA

NICARAGUA

PANAMA

◀ **Tropical forests**
Large areas of the lush tropical forests which once covered Central America have been destroyed by loggers and farmers.

◀ **Temple of the Warriors**
This imposing ruin was built by the fierce Toltec warriors who conquered the Maya of Chichén Itzá in AD987.

Map labels

Tijuana, Mexicali, Ensenada, Ciudad Juárez, Hermosillo, Chihuahua, Cedros I., Gulf of California, Baja California, SIERRA MADRE, Rio Bravo del Norte, Rio Grande, UNITED STATES OF AMERICA, Torreón, Saltillo, Monterrey, Matamoros, La Paz, Culiacán, Durango, San Luis Potosí, Tampico, Aguascalientes, Guadalajara, León, Cape Corrientes, L. de Chapala, Manzanillo, **Mexico City**, Bay of Campeche, Campeche, Mérida, Cancún, Yucatán Peninsula, Yucatán Channel, **MEXICO**, Puebla, Orizaba 5,700 m, Veracruz, Terminos Lagoon, Villahermosa, Belize City, Belmopan, Acapulco, Balsas, Coatzacoalcos, Oaxaca, **BELIZE**, PACIFIC OCEAN, Gulf of Tehuantepec, **GUATEMALA**, **HONDURAS**, Guatemala City, Tegucigalpa, San Salvador, **EL SALVADOR**, **NICARAGUA**, Managua, Lake Nicaragua, Mosquitos Gulf, **PANAMA**, San José, **COSTA RICA**, Panama City, COLOMBIA, Gulf of Panama

GUATEMALA   HONDURAS

NICARAGUA

EL SALVADOR

COSTA RICA

PANAMA

Mexico and Central America

MEXICO IS NORMALLY CONSIDERED AS part of North America, although its southern regions have more in common with the Central American nations which lie to the south. It is a federal democratic republic and in 1994 signed up to the North American Free Trade Agreement (NAFTA) with Canada and the USA.

Mexico has large oilfields in the Gulf of Mexico as well as silver, lead, gold and uranium. Ninety-five percent of its mineral resources are still to be mined. It manufactures fertilizers, petrochemicals, vehicles and machines. Mexican farmers grow cotton, coffee, tropical fruits and vegetables. There is a large fishing industry.

Even so, Mexico faces many problems. It has the highest foreign debt of any developing country. Many of its people are very poor; over the years large numbers of them have crossed illegally into the United States in search of work. The population of Mexico City has soared above 20 million and it is probably the most polluted capital city in the world. In 1994 poor peasants in the southern state of Chiapas rose in rebellion against the central government.

▲ Coffee beans
Coffee is a major export of Central American countries such as Nicaragua and Costa Rica.

▲ Blue waters
Off Cajún, in northeast Yucatán, glass-bottomed boats allow tourists to see the wonders of a tropical coral reef.

▲ Sea harvest
A Mexican fisherman casts his net. Sardines are by far the largest catch in these waters, followed by anchovies, tuna and prawns.

▼ Down in Acapulco
Tourists flock to Acapulco in southwestern Mexico for the beaches and fishing. The city has one of the best natural harbours in the Pacific.

▲ Guacamole dip, tacos and chilli peppers

FACTS

HONDURAS
República de Honduras
Area: 112,088 sq km
Population: 6.1 million
Capital: Tegucigalpa
Other cities: San Pedro Sula, Choluteca
Highest point: Cerro Las Minas (2,849m)
Official language: Spanish
Currency: lempira

MEXICO
Estados Unidos Mexicanos
Area: 1,958,201 sq km
Population: 93.2 million
Capital: Mexico City
Other cities: Guadalajara, Monterrey, Puebla de Zaragoza
Highest point: Citlatépetl (5,700m)
Official language: Spanish
Currency: Mexican peso

NICARAGUA
República de Nicaragua
Area: 130,000 sq km
Population: 4.5 million
Capital: Managua
Other cities: León, Granada
Highest point: Cordillera Isabella (2,438m)
Official language: Spanish
Currency: córdoba

PANAMA
República de Panamá
Area: 75,517 sq km
Population: 2.7 million
Capital: Panama City
Other cities: San Miguelito, Colón, David
Highest point: Volcán Baru (3,475m)
Official language: Spanish
Currency: balboa

◄ Mexico City
The Mexican capital and its sprawling suburbs are home to nearly 25 million people. It is the most populous city in all the Americas and is a centre of business, finance, industry and communications.

▶ Tropical produce
The tropical climate produces a variety of crops, with sugarcane grown in the humid lowlands and potatoes cultivated on cooler mountain slopes.

Mexico's problems are mirrored in the Central American republics of Guatemala, Belize, Honduras, El Salvador, Nicaragua, Costa Rica and Panama. Over the last 50 years, Central America suffered from extreme right-wing dictators and political parties, who have often tried to silence all opposition with death squads. These governments were sometimes backed by the United States, who wanted to prevent the region from becoming communist. The governments came under prolonged attack from left-wing revolutionaries. Central America also saw bitter disputes over territories and borders.

The root of all these conflicts has been grinding poverty. The warfare has now mostly ended, but the rebuilding process is very slow. Poverty and foreign debt remain the chief problem. The regional economy depends on coffee, bananas, cotton, sugarcane, maize, fish products and seafood. Exports include textiles and handicrafts.

▲ Banana trade
Bananas are packed for export before they ripen. The Central American countries compete fiercely with their Caribbean neighbours for their share of the world market.

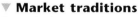
▼ Market traditions
Bargain-hunters throng an outdoor flea market in Mexico City. The modern city is built on the site of the ancient Aztec capital, Tenochtitlán, which was famous for its vast, open-air markets.

▲ Woven by hand
The region is famous for its beautifully coloured and patterned textiles. Traditionally, these are woven by hand, using simple backstrap looms.

▶ Mining for tin
Mexico has rich mineral reserves, including tin, silver, antimony, mercury and fluorite. Its most valuable reserves are of oil and natural gas.

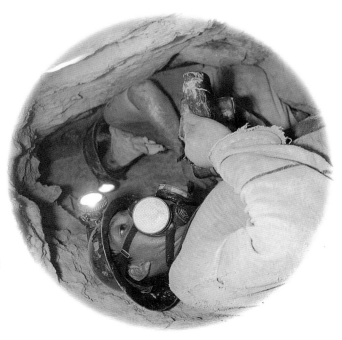

Mexico and Central America

FROM ABOUT 1200BC ONWARDS, MEXICO AND Central America saw the rise of many great civilizations founded by indigenous (native) peoples such as the Olmecs, Maya, Toltecs and Aztecs. Ruined cities, great pyramids and temples may still be seen. When the Spanish invaded the region in AD1519, they were awestruck by the wonders of the Aztec capital, Tenochtitlán (today's Mexico City).

The indigenous peoples, whom the Spanish called 'Indians', were skilled astronomers, mathematicians, writers, musicians and craftspeople. However they had no firearms and so were soon defeated. The region became part of Spain's overseas empire for the next three hundred years or so.

After independence in 1823, power remained in the hands of a few wealthy landowners. Mexico lost large areas of territory to the United States in the 1840s. From 1910 until 1917 the country was torn apart by revolution and civil war.

▲ **An Olmec head**
This colossal stone head was carved about three thousand years ago by the Olmec people of ancient Mexico.

▲ **An Aztec headdress**
The Aztecs were fine craftworkers, producing beautiful feather cloaks and headdresses, intricate jewellery and finely-patterned textiles.

▼ **Colonial architecture**
Spanish-built churches may still be seen in Taxco de Alarcón, Mexico. The region was already a centre of mining before the Spanish came here in 1528.

▲ **Mexican mummies**
Peoples of ancient Mexico, such as the Aztecs and Mixtecs, bundled up mummies of the dead. These might be burned or buried.

◄ **Mitla ruins**
Mitla, in southern Mexico, was an ancient holy site. It was occupied by the Mixtecs about a thousand years ago. Its ruins still bear traces of the original paint, made from berries.

▼ **The Maya**
The classic period of Mayan civilization lasted from about AD250 to 900. It was marked by the building of cities with great stone temples and pyramids. The homeland of the Maya stretched from Mexico's Yucatán peninsula southwards into Guatemala and Belize. Their descendants still live in these regions today.

◄ Textile weaving
Skeins of brilliant coloured wools are piled up high ready for sale in this Guatemalan market place.

▼ Fiesta!
Festivals are an important part of both Spanish and local traditions in Mexico.

◄ Violinist
Mariachi music originated in Jalisco, Mexico. Sentimental songs are played on violins, guitars and trumpets.

After the Central American nations gained independence, they at first tried to unite in a federation. However they went their separate ways in 1838. Here too, the people remained desperately poor while a few landowners grew very rich. Belize was a British colony called British Honduras from 1862 until 1981.

Despite a troubled history, the peoples of the region enjoy a rich culture. Some are of European descent, but these are outnumbered by mestizos, people of mixed native and Spanish descent. Indigenous peoples still living in the region include large groups of Maya, as well as Otomi, Tarascan, Zapotec, Mixtec, Tarahumara, Nahua, Miskito, Guaymi and Cuna. The Garifuna, who live in Belize, Honduras and Nicaragua, are descended from Africans and from an indigenous people called the Caribs. The whole region is Spanish-speaking. Indigenous languages are also spoken and English is heard in Belize and on Nicaragua's east coast.

Mexico and Central America are strongly Roman Catholic, although indigenous traditions have influenced many colourful religious festivals and processions.

▲ Day of the Dead
Mexicans commemorate their dead each year on November 1st. Altars are laden with food offerings, papier-mâché skeletons, photographs and flowers.

▼ Mixed traditions
Ancient Mayan rituals have influenced the Christian worship of the Quiché Maya, of Guatemala.

TIMELINE

BC	
c.2600	Origins of Maya civilization in Yucatán
c.1000	Origins of Zapotec civilization
AD	
300s	Mayan empire (until 900s)
c.1325	Aztecs build great city of Tenochtitlán
1400s	Height of Aztec empire
1519	Spain invades Aztec empire (conquers Aztecs by 1521 and goes on to colonize all Central America)
1600s	Pirate attacks on Atlantic coast
1810	Revolt against Spanish rule in Mexico
1821	Collapse of Spanish rule in Central America
1823	United provinces of Central America (by 1838): Honduras, Costa Rica, El Salvador, Guatemala, Nicaragua
1846	Mexican War with USA (until 1848): loss of California and New Mexico
1862	British Honduras (Belize) a British colony
1911	Mexican Revolution: political reform
1914	Panama Canal opens
1970s	Political violence in Guatemala
1978	Nicaraguan Revolution
1981	Belize independent
1985	Earthquake, Mexico City
1994	Uprising in Chiapas, Mexico
	Mexico enters North American Free Trade Agreement (NAFTA)
1998	Hurricane Mitch

121

The Caribbean

Many beautiful islands lie between the Straits of Florida and the Venezuelan coast of South America. They are bordered in the west by the Gulf of Mexico and in the east by the open Atlantic. Many of the islands are formed from coral or from volcanic rock. Some have active volcanoes. Offshore reefs form spectacular underwater worlds of corals, seaweeds and brilliantly-coloured fish.

The climate is generally tropical and warm, with the humidity relieved by ocean breezes on the smaller islands. Fierce storms called hurricanes bring high winds and rains in late summer and autumn.

The Bahamas and the Turks and Caicos islands form the northern, outer ring of islands. The Greater Antilles chain includes Cuba, the largest island of the region. It has fertile plains, rolling hills, forests and mountains. Its neighbours are the Cayman Islands, Jamaica, Hispaniola (divided between Haiti and the Dominican Republic) and Puerto Rico.

The islands of the Lesser Antilles are smaller, scattered around a great arc between the Virgin Islands and Aruba. The northern ones are known as the Leeward Islands and the southern ones as the Windward Islands. The area enclosed by the Antilles is called the Caribbean Sea.

▶ **Scarlet ibis**

▲ **Donkey traffic**
Donkeys are a traditional way of getting around in the mountainous countryside of the Dominican Republic.

FACTS

ANTIGUA AND BARBUDA
Area: 442 sq km
Population: 64,000
Capital: St Johns
Highest point: Boggy Peak (402 m)
Official language: English
Currency: East Caribbean dollar

BAHAMAS
Commonwealth of the Bahamas
Area: 13,878 sq km
Population: 0.27 million
Capital: Nassau
Highest point: Cat Island (63 m)
Official language: English
Currency: Bahamian dollar

BARBADOS
Area: 430 sq km
Population: 0.26 million
Capital: Bridgetown
Highest point: Mt Hillaby (340 m)
Official language: English
Currency: Barbados dollar

CUBA
República de Cuba
Area: 110,861 sq km
Population: 11.0 million
Capital: Havana
Other cities: Santiago de Cuba, Camagüey
Highest point: Turquino (2,005 m)
Official language: Spanish
Currency: Cuban peso

▼ Stormy weather

Satellites in space track the progress of a hurricane. These devastating, whirling storms sweep across the Caribbean between August and October each year.

▲ Palm trees

Everyone's image of the Caribbean is of beautiful palm trees swaying in the breeze. Here, on the sunny island of Jamaica, coconuts are growing in trees that can be as tall as 24 metres.

CUBA

BAHAMAS

PUERTO RICO

ANTIGUA AND BARBUDA

ST KITTS AND NEVIS

DOMINICA

ST LUCIA

BAHAMAS

•Nassau

Andros I.

Turks & Caicos Islands (U.K.)

MEXICO

Havana

CUBA

Camagüey

Santiago de Cuba

Isla de la Juventad

Cayman Islands (U.K.)

Port-au-Prince

Kingston

JAMAICA

HAITI

DOMINICAN REPUBLIC

Santo Domingo

San Juan

Puerto Rico (U.S.)

Virgin Is. (U.K. & U.S.)

ST. KITTS & NEVIS

ANTIGUA & BARBUDA

Montserrat (U.K.)

Guadeloupe (FR.)

DOMINICA

Martinique (FR.)

ST. LUCIA

BARBADOS

ST. VINCENT & THE GRENADINES

GRENADA

Netherlands Antilles

TRINIDAD & TOBAGO

GREATER ANTILLES

CARIBBEAN SEA

LESSER ANTILLES

JAMAICA

HAITI

DOMINICAN REPUBLIC

TRINIDAD AND TOBAGO

GRENADA

ST VINCENT AND THE GRENADINES

BARBADOS

▼ Coral reef

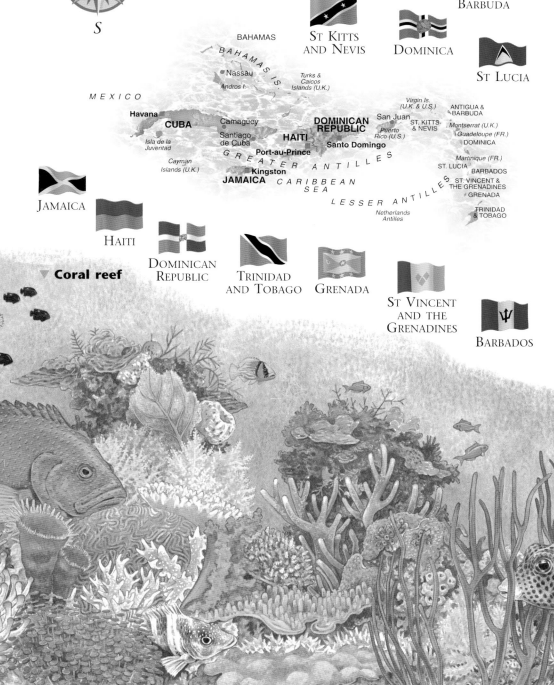

FACTS

DOMINICA
Commonwealth of Dominica
Area: *751 sq km*
Population: *0.07 million*
Capital: *Roseau*
Highest point: *Morne Diablotin (1,447 m)*
Official language: *English*
Currency: *East Caribbean dollar*

DOMINICAN REPUBLIC
República Dominicana
Area: *48,734 sq km*
Population: *8.0 million*
Capital: *Santo Domingo*
Other cities: *Santiago de los Caballeros, La Romana*
Highest point: *Pico Duarte (3,175 m)*
Official language: *Spanish*
Currency: *Dominican Republic peso*

GRENADA
Area: *344 sq km*
Population: *0.1 million*
Capital: *St George's*
Highest point: *Mt St Catherine (840 m)*
Official language: *English*
Currency: *East Caribbean dollar*

HAITI
République d'Haïti
Area: *27,750 sq km*
Population: *7.3 million*
Capital: *Port-au-Prince*
Other cities: *Jacmel, Les Cayes*
Highest point: *La Selle (2,677 m)*
Official languages: *French, Creole*
Currency: *gourde*

The Caribbean

THERE ARE 13 INDEPENDENT NATIONS IN the Caribbean region. Some are democratic republics, some still have European monarchs as their heads of state. Economic links are strengthened by the Caribbean Community and Common Market (Caricom, founded in 1973). Most of the Caribbean's 11 other territories remain dependencies of other countries by choice, because it brings them wealth or security. Some are governed as if they were part of mainland France.

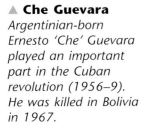

▲ Che Guevara
Argentinian-born Ernesto 'Che' Guevara played an important part in the Cuban revolution (1956–9). He was killed in Bolivia in 1967.

Cuba, the largest Caribbean island, has had a communist government since 1959. This has been bitterly opposed by the United States. Until 1991, Cuba had vital trading links with the communist countries of Central and Eastern Europe. Since then, Cuba has had to struggle against a strict United States ban on trade.

Political violence and corruption have a long history on some islands, such as Haiti. On other islands the problem has been one of alliances. For example, Anguilla refused to join the federation of St Kitts–Nevis when it became independent, and Aruba pulled out of the Netherlands Antilles. Other islands have been hard hit by natural disasters, such as hurricane damage or volcanic eruptions.

▲ Fidel Castro
The leader of Cuba's revolution was Fidel Castro. He overthrew the corrupt regime of Fulgencio Batista in 1959 and was still Cuban president 40 years later.

▲ World-class cigars
Cuba has long been famous for the finest quality cigars and for rum, made from sugarcane.

▲ Radio telescope
The world's biggest single-unit radio telescope is located near Arecibo, on the island of Puerto Rico.

◄ Boat building
A wooden hull is constructed in Barbados. Boats are still built by traditional methods on many Caribbean islands.

FACTS

JAMAICA
Area: *10,990 sq km*
Population: *2.5 million*
Capital: *Kingston*
Other cities: *Spanish Town, Montego Bay*
Highest point: *Blue Mountain Peak (2,256 m)*
Official language: *English*
Currency: *Jamaican dollar*

ST KITTS – NEVIS
Federation of St Christopher and Nevis
Area: *261 sq km*
Population: *0.04 million*
Capital: *Basseterre*
Highest point: *Liamuiga (1,156 m)*
Official language: *English*
Currency: *East Caribbean dollar*

ST LUCIA
Area: *622 sq km*
Population: *0.16 million*
Capital: *Castries*
Highest point: *Mt Gimie (959 m)*
Official language: *English*
Currency: *East Caribbean dollar*

ST VINCENT AND THE GRENADINES
Area: *388 sq km*
Population: *0.11 million*
Capital: *Kingstown*
Highest point: *La Soufrière (1,272 m)*
Official language: *English*
Currency: *East Caribbean dollar*

TRINIDAD AND TOBAGO
Republic of Trinidad and Tobago
Area: *5,130 sq km*
Population: *1.3 million*
Capital: *Port-of-Spain*
Other cities: *San Fernando, Arima*
Highest point: *Aripo (940 m)*
Official language: *English*
Currency: *Trinidad and Tobago dollar*

◀ Market day
What's on sale in a Caribbean market? Lush tropical fruits, coconuts, saltfish, seafood and vegetables. Snacks on offer are sure to include rotis – pancakes wrapped around curried chicken and vegetables.

▼ Changing times
St Lucia's Rodney Bay was once a swampy coastline with a couple of fishing villages. In the 1940s, it was transformed into a US naval airbase. Today, it is a tourist centre, with a yachting marina and many hotels.

Many of the Caribbean islanders are poor. Cash crops, produced for export, include spices, tropical fruits such as bananas, mangoes and limes, sugarcane and cotton. Cuba is famous for its rum and tobacco. Many islanders fish and grow their own food. Popular local dishes are made from saltfish, pigeon peas, coconut, chilli peppers, and cornmeal.

Manufacturing is limited, but Trinidad produces oil and natural gas. The region's blue seas and beaches of white sand fringed with palms have made tourism a major industry in many islands. Other islands, such as the Caymans, have passed tax laws which allow international finance companies and banks to set up their headquarters there. The wealth made from tourism and offshore banking often fails to benefit the local people.

▲ Sailfish
With a pointed bill and a sail-like fin, this powerful fish can reach speeds of up to 80 kilometres per hour.

▼ Caribbean village
Most Caribbeans live in simple, one-storey homes. White walls reflect the sun and keep the interiors cool despite the fierce heat outside. Up to four generations of the same family may live together in one or these small houses.

FACTS

Caribbean dependencies

ANGUILLA
British Overseas Territory
Area: *91 sq km*
Population: *9,000*
Capital: *The Valley*
Official language: *English*
Currency: *East Caribbean dollar*

ARUBA
Self-governing island of the Netherlands
Area: *193 sq km*
Population: *0.08 million*
Capital: *Oranjestad*
Official language: *Dutch*
Currency: *Aruban guilder*

CAYMAN ISLANDS
British Overseas Territory
Area: *264 sq km*
Population: *0.03 million*
Capital: *George Town*
Official language: *English*
Currency: *Cayman Islands dollar*

GUADELOUPE
Overseas region of France
Area: *1,705 sq km*
Population: *0.4 million*
Capital: *Basse-Terre*
Official language: *French*
Currency: *French franc*

MARTINIQUE
Overseas region of France
Area: *1,102 sq km*
Population: *0.4 million*
Capital: *Fort-de-France*
Official language: *French*
Currency: *French franc*

The Caribbean

IN PREHISTORIC TIMES THE CARIBBEAN ISLANDS WERE settled by indigenous peoples from the American mainland. When Columbus discovered the islands in 1492, these formed two main groups – Arawak-speaking peoples and Caribs. The Arawaks were savagely treated by the invaders. They were enslaved, murdered or infected with diseases. The Caribs, who had a fearsome reputation amongst the Europeans, resisted longer. In the end, however, few communities survived.

The Spanish took over many islands. As their fleets shipped back plundered treasures from the mainland (the 'Spanish Main'), they were preyed on by British, French and Dutch pirates. Soon these countries too seized islands and planted them with tobacco or sugarcane. They imported slaves from Africa to work the land. Cruelly treated, some slaves escaped and there were violent revolts in Jamaica and Haiti in the 1700s. In the 1800s, Haiti became the first independent Afro-Caribbean republic.

▲ The *Santa Maria*
Christopher Columbus' ship reached San Salvador, in the Bahamas, in 1492 – a turning point in American history.

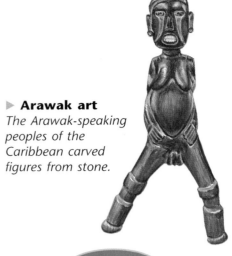

▶ Arawak art
The Arawak-speaking peoples of the Caribbean carved figures from stone.

▲ Sugar plantations
Sugarcane dominated the Caribbean economy from the 1600s until recent times. Sugar was shipped to Europe and North America. The backbreaking labour on the European-owned plantations was carried out by slaves imported from West Africa.

▼ The slave trade
Slavery was the curse of the Caribbean from the 1500s until its abolition three centuries later. Slaves were sold like cattle, families were separated and punishments were harsh, often fatal.

◀ Haitian voodoo
Drums and dancing send people into a trance at Cap Haitien. Voodoo, known as Vodun in the Creole dialect of Haiti, is based on an African belief in spirits called loas. It inspired the slave revolts on Haiti and is still popular today.

FACTS

MONTSERRAT
British Overseas Territory
Area: *102 sq km*
Population: *0.01 million (6,000 since volcanic eruptions of 1995–7)*
Capital: *Plymouth (evacuated)*
Official language: *English*
Currency: *East Caribbean dollar*

NETHERLANDS ANTILLES
Self-governing islands of the Netherlands
Area: *800 sq km*
Population: *0.2 million*
Capital: *Willemstad*
Official language: *Dutch*
Currency: *Netherlands Antilles guilder*

PUERTO RICO
Commonwealth territory of the USA
Area: *8,875 sq km*
Population: *3.8 million*
Capital: *San Juan*
Official languages: *English, Spanish*
Currency: *US dollar*

TURKS AND CAICOS ISLANDS
British Overseas Territory
Area: *430 sq km*
Population: *0.01 million*
Capital: *Cockburn Town*
Official language: *English*
Currency: *US dollar*

VIRGIN ISLANDS (BRITISH)
British Overseas Territory
Area: *151 sq km*
Population: *0.02 million*
Capital: *Road Town*
Official language: *English*
Currency: *US dollar*

VIRGIN ISLANDS (USA)
Unincorporated Territory of the USA
Area: *347 sq km*
Population: *0.1 million*
Capital: *Charlotte Amalie*
Official language: *English*
Currency: *US dollar*

▼ Steel band
Steel drums
called pans
provide the typical
sound of Trinidad.

► Carnival
Carnival in
Trinidad is a time
of glitzy costumes,
dancing, steel-
band music and
verses called
calypsos.

▲ Pirate women
Two notorious women
pirates, Anne Bonny and
Mary Read, went on trial
in Jamaica in 1720.
Caribbean coasts had long
been terrorized by pirates
and buccaneers.

After the abolition of slavery (mostly in the 1830s) many
European settlers left the region they called the West Indies.
At the same time labourers from India and Southeast Asia
were hired to work on some islands, such as Trinidad.
Most Caribbean islands remained colonies until the 1960s.

Today's Caribbeans are descended from many roots – Native
American, African, Spanish, English, Irish, French, Dutch and
Asian. Much the largest ethnic groups are of Afro-Caribbean
or Hispanic descent. Spanish, English and French are widely
spoken, often in strong Creole dialects influenced by African
languages. The region is strongly Christian, with some followers
of African spirit religions, such as the
Voodoo worship on Haiti. The
Rastafarians of Jamaica are also
inspired by African spirituality.

The Caribbean's mixed roots have
inspired a range of popular music
and dance styles that have become
popular all over the world, from
salsa to reggae. Music comes to
the fore in the region's famous
carnivals. Popular sports
include cricket, soccer
and baseball.

▲ Cricket
West Indian cricket
teams have been
world-beaters since
the 1960s.

◄ Back to Africa
Rastafarians look back to
their African roots and
honour Haile Selassie
(1891–1975), former
emperor of Ethiopia.

AD	TIMELINE
c.200	Native American settlement: Ciboney people expelled from Cuba by the Taíno
c.1000	Arawak and Carib migrations and wars
1492	Christopher Columbus lands in the Bahamas
1496	First Spanish settlement, Santo Domingo (Hispaniola)
1511	Spanish settle Cuba
1523	Slave trade with Africa
1655	British capture Jamaica
1660s	Piracy widespread (until 1720s)
1697	French gain control of Haiti (Hispaniola)
1804	Haiti independent republic under black rule
1838	Abolition of slave trade in some parts of the region
1868	Independence movement defeated in Cuba (by 1878)
1895	Uprising in Cuba (until 1898)
1898	Puerto Rico ceded to USA
1952	Fulgencio Batista seizes power in Cuba
1959	Fidel Castro overthrows Batista in Cuba
1961	Barbados independent
1962	Cuban Missile Crisis: clash with USA over siting of Soviet missile base Jamaica, Trinidad independent
1973	Bahamas independent
1983	USA invades Grenada
1995	Series of volcanic eruptions begins on Montserrat

Northern Andean countries

▲ **Condor**

The Andes mountains run down the whole length of South America, from Colombia to Tierra del Fuego, a distance of over 6,400 kilometres. The mountains are very high, very beautiful and very valuable, for they contain precious silver, tin and other minerals.

The range includes massive snow-capped volcanoes, gleaming glaciers and wide, cool plateaus. Titicaca, the world's highest navigable lake, occupies 8,288 square kilometres of the plateau on the border between Peru and Bolivia.

To the north, the Andes drop to low-lying plains around the humid Caribbean coast. To the west, they are bordered by a coastal strip along the Pacific Ocean. In parts this is humid and fertile, but in Peru much of it is dusty desert. That is because cold ocean currents make it more difficult for the air to fill with moisture and form rain. Far to the west, Ecuador also takes in a Pacific island chain, the Galápagos.

▲ **Bird and snake**
Beautiful stone carvings were produced at San Augustín, Colombia, about 2,000 years ago.

To the east, the peaks and sunlit plateaus of the Andes descend through misty foothills and sheer-sided valleys into the vast rainforests of central South America. Streams rising on the eastern slopes drain into the great Orinoco and Amazon river systems.

▼ **Buttress roots**

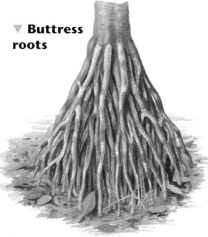

▲ **One more river**
The Urubamba rises in the Andes and flows through deep gorges. It eventually joins up with the Apurímac to form the Ucayali, which in turn drains into the mighty Amazon.

FACTS

BOLIVIA
República de Bolivia
Area: *1,098,581 sq km*
Population: *7.6 million*
Capitals: *La Paz, Sucre*
Other cities: *Santa Cruz, Cochabamba*
Highest point: *Nevado Sajama (6,542 m)*
Official language: *Spanish*
Currency: *peso boliviano*

COLOMBIA
República de Colombia
Area: *1,138,914 sq km*
Population: *37.4 million*
Capital: *Bogotá*
Other cities: *Medellín, Cali, Barranquilla*
Highest point: *Cristobal Colón (5,775 m)*
Official language: *Spanish*
Currency: *Colombian peso*

ECUADOR
República del Ecuador
Area: *461,475 sq km (including Galapágos Islands)*
Population: *11.7 million*
Capital: *Quito*
Other cities: *Guayaquil, Cuenca*
Highest point: *Chimborazo (6,267 m)*
Official language: *Spanish*
Currency: *sucre*

PERU
República de Peru
Area: *1,285,216 sq km*
Population: *24.3 million*
Capital: *Lima*
Other cities: *Callao, Arequipa, Chiclayo*
Highest point: *Nevado Huascarán (6,768 m)*
Official languages: *Spanish, Quechua, Aymará*
Currency: *nuevo sol*

▶ Andes watershed
South America is divided into two by the Andes range. Some rainfall drains into the Pacific Ocean, but the rest flows eastwards into the distant Atlantic.

▲ Baños, Ecuador
This small town lies in the eastern Andes, on the route to El Oriente, Ecuador's province in the Amazon River basin.

◀ River of ice
In Peru, the Andes divide into three main sections. The highest peaks rise in the Cordillera Blanca. Here, snow-capped peaks tower above glaciers such as Pastoruri, at 5,300 metres above sea level.

COLOMBIA

ECUADOR

Point. Gallinas
Barranquilla
Cartagena
Cristobal Colón 5,775 m
PANAMA
VENEZUELA
Cauca
Magdalena
Cape Corrientes
Medellín
Pereira
Manizales
Meta
Ibague
Bogotá
COLOMBIA
Buenaventura
Cali
Neiva
Nevado del Huila 5,750 m
Pasto
Guaviare
Point Galera
Quito
ECUADOR
Caquetá
Guayaquil
Chimborazo 6,267 m
Putumayo
Gulf of Guayaquil
Amazon
Marañón
Iquitos
Point Aguja
Piura
Chiclayo
BRAZIL
Ucayali
Trujillo
Chimbote
Nevado Huascarán 6,768 m
PERU
Callao
Huancayo
Lima
Paracas Pen.
Cuzco
Nazca
Volcán El Misti 5,842 m
Nevado Ancohume 6,550 m
Arequipa
Lake Titicaca
La Paz
BOLIVIA
Mamoré
Guaporé
PACIFIC OCEAN
Cochabamba
Santa Cruz
Oruro
Lake Poopó
Sucre
Potosí
Pilcomayo
CHILE
ALTIPLANO
PARAGUAY

N
W E
S

PERU

BOLIVIA

▲ Eastern Bolivia
This hamlet lies in the Las Yungas region of eastern Bolivia, where the Altiplano (high plateau) drops to the humid, forested lowlands of the Amazon basin.

▼ Lake Titicaca
This is the largest lake in South America. It is really made up of two smaller lakes, Chucuito and Uinamarca, which are linked by a narrow strait.

▶ Llama
For thousands of years the llama has provided wool and meat, as well as being used to carry loads over high Andean passes.

◀ High altitude
The Chacaltuya ski lodge near La Paz, Bolivia, is the world's highest. It overlooks the tablelands of the Altiplano.

129

Northern Andean countries

THE FOUR COUNTRIES OF THE NORTHERN ANDES are all republics. The army has often seized power in this region, or controlled governments behind the scenes. There have been long years of guerrilla warfare, fuelled by widespread poverty and social injustice. There have been murderous activities by the criminal gangs who export cocaine. This illegal drug is made from the coca plant, which has been grown by poor peasants in the region for thousands of years.

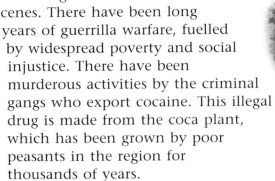

▲ **Cocaine patrol**
A plane hunts for secret coca plantations. This plant, used as a drug for thousands of years, is processed to make cocaine.

▲ **Mule train**
Surefooted but stubborn, mules have been used to transport goods in the Andes since the 1500s.

The northern Andes region has great mineral wealth in the form of oil, copper, emeralds, silver, tin, zinc, lead, silver and gold. However many people remain very poor, in both city and country areas. In the hot, lowlands, farmers grow cotton, sugarcane and bananas. Colombian coffee beans, grown on tropical slopes, are among the best in the world. On the the cool plateaus, potatoes, maize, wheat and a grain called quinoa are grown.

The forests produce valuable timber. Sardines, anchovies and tuna are caught by the large Pacific fishing fleets. Shrimps are farmed along the coast. Fish are also processed to make fertilizers.

▲ **Farming the slopes**
Farmers were already terracing the steep Andean slopes in ancient times, in order to conserve soil. The Incas were masters of irrigation techniques.

◀ **Upriver**
Motorized canoes travel along the Río Aguarico, which and flows through eastern Ecuador.

▲ **The fur trade**
In Colombia, illegal hunting claims the lives of many rare wildlife species, such as the ocelot.

▲ **Silver mountain**
When the Spanish conquered Bolivia, Cerro Rico, in Potosi, became the world's biggest silver mine. The riches were shipped back to Europe.

▼ **Andean station**
Alausí, a centre of population in Ecuador's Chimborazo province, is built on the Guayaquil–Quito railway. The world's highest railways are in the Andes, in Peru.

◄ Lima stadium
Peru's National Stadium is sited in Lima, the capital. Lima is home to nearly six million people.

The region is strongly Roman Catholic and there are many religious festivals and pilgrimages. Spanish is spoken in all four countries. There are many people of mestizo (mixed European and Indian) as well as a number of European, African or Asian descent. Indigenous (native) peoples make up just one percent of the population in Colombia, but 25 percent in Ecuador, 45 percent in Peru, and 55 percent in Bolivia.

The two largest indigenous groups are the Aymara and Quechua. Most of these 'Indians' are poor peasants and are largely ignored by those who hold power in the cities. In the mountains, life carries on as it has for centuries, with the local people weaving, going to market and herding llamas. The haunting folk music of the Andes is played on guitars, panpipes, flutes and drums.

▲ Local transport
A motorized rickshaw provides transport in Chiclayo, the chief town of Peru's Lambayeque region. It lies on the Pan-American highway, the road network which links North to South America.

▼ Columbian capital
Mountains rise behind Bogotá's modern high-rise skyline. Colombia's chief city is home to some five million people.

◄ Straw hats
Ecuador manufactures panama hats. These are woven from the leaves of a palm-like shrub called jipyapa.

► Bamboo pipes
Known as antaras or zampoñas, panpipes have been played since the days of the Incas.

► Fishing boats
The Aymara and Uru peoples of Lake Titicaca make fishing boats called balsas from reeds. Fish caught in this deep lake include catfish, trout and killifish.

◄ Gathering reeds
Tall, tough reeds called totora grow on the shores and islands of Lake Titicaca and on floating platforms of vegetation. They are harvested and made into houses as well as fishing boats.

131

Northern Andean countries

DESPITE ITS JAGGED MOUNTAIN PEAKS, VOLCANOES and deserts, the northern Andes region was settled by the waves of tribes who peopled the Americas in prehistoric times. It became the site of many very advanced ancient civilizations.

From about 1200BC until 200BC the Chavín civilization produced stone carvings, pottery and painted textiles. Strange lines scratched deep into the desert near Nazca in southern Peru may have been designed to honour the gods. Further north, the Moche valley was the centre of a civilization which built pyramids in honour of the Sun, and engineered an irrigation system for crops in dry regions.

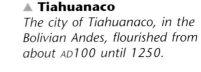

▲ **Tiahuanaco**
The city of Tiahuanaco, in the Bolivian Andes, flourished from about AD100 until 1250.

▲ **Inca stones**
The ancient Incas mastered the skills of building in stone. Their buildings have withstood the strongest earthquakes.

Over 1,000 years ago, the Chimú people of northern Peru were masters of working gold and they also produced pottery on a large scale. The most amazing civilization of all was that of the Incas, who defeated the Chimú in AD1476. The Inca empire, with its capital at Cuzco in Peru, stretched from Ecuador to Chile and took in about 12 million people. The Incas were builders of cities and towns, who studied the stars and built great temples.

▶ **Machu Picchu**
This lost city of the Incas, high above the gorges of the Urubamba, was rediscovered in 1911.

▲ **Peruvian funeral mask**

▶ **Pilgrimage**
The Spanish conquerors brought Roman Catholicism to the Andes. This is the Sanctuary of Nuestra Señora de Acua Santa, in Ecuador.

▲ **Temple of the Sun**
This pyramid dominated the Moche capital. Forty metres high, it is believed to contain about 140 million mud bricks.

▶ **Conquest**
The Spanish reached Peru in 1532. They captured the emperor, Atahualpa, by treachery and soon conquered all the Inca lands, although resistance continued for 40 years.

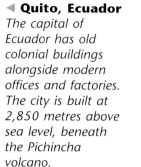

Quito, Ecuador
The capital of Ecuador has old colonial buildings alongside modern offices and factories. The city is built at 2,850 metres above sea level, beneath the Pichincha volcano.

La Paz, Bolivia
A street vendor sells vegetables in La Paz, one of Bolivia's twin capitals. La Paz is the world's highest capital, at 3,631 metres.

Spain invaded Colombia in 1499. By 1532 a small force of Spanish soldiers, called conquistadors, had used treachery to overthrow the powerful Inca empire to the south. They were greedy for gold and silver, and the northern Andes offered wealth beyond their dreams.

Lure of gold
The Spanish were first attracted to South America by rumours of fabulous gold.

Spain's rule over the region lasted nearly 300 years. They built churches and cities, mined for silver and used the indigenous people to labour on the land. In the 1820s a revolutionary called Simón Bolívar (1783–1830) led the struggle against continuing rule by Spain. His name was taken up by one of the territories he freed, Bolivia.

The newly-independent countries were soon attacking each other. Wealth remained within a few families of the ruling class, while most people remained wretchedly poor into modern times.

Mountain dress
Many Bolivians still wear traditional dress, including blankets, shawls and these distinctive bowler hats.

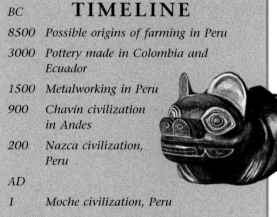

Inca trail
The Inca empire was linked by a vast network of roads, bridges and mountain trails. Goods were carried by llama. Relays of runners carried messages by hand.

TIMELINE

BC

8500	Possible origins of farming in Peru
3000	Pottery made in Colombia and Ecuador
1500	Metalworking in Peru
900	Chavín civilization in Andes
200	Nazca civilization, Peru

AD

1	Moche civilization, Peru
600	Tiahuanaco civilization, Bolivia
1200	Beginnings of Inca empire
1250	Chimu civilization, Peru
1400s	Inca empire at greatest extent
1531	Spanish land in Ecuador
1533	Incas defeated by Spanish
1538	Colombia conquered by Spanish
1545	Silver mining in Bolivia
1780	Indigenous uprising in Peru
1819	Colombia independent from Spain
1822	Ecuador joins Colombia
1824	Peru independent from Spain
1825	Bolivia independent from Spain
1830	Full independence for Colombia Ecuador independent
1884	Bolivia loses Atacama to Chile
1932	Chaco War (until 1935): Bolivia defeated by Paraguay
1941	Peru invades Ecuador
1948	OAS (Organization of American States) is signed at Bogotá, Colombia
1952	Bolivian National Revolution
1989	Drug gang violence in Colombia

133

Brazil and its neighbours

◄ Poison-arrow frog

Venezuela ('Little Venice') was given its name because the vast, shallow inlet of Lake Maracaibo reminded early European explorers of the lagoons around Venice, in Italy.

Venezuela is dominated by the river Orinoco, which forces its way between a northern spur of the Andes Mountains and the Guiana Highlands. It flows through the grassy plains of the Llanos to form a wide, swampy delta on the Caribbean coast. In the central southeast, the Angel Falls form the world's highest waterfall, with a total drop of 979 metres.

The low-lying Caribbean coast, with a hot and sticky climate, continues eastwards through Guyana, Surinam and French Guiana. In the south of these countries the land rises to the forested slopes of the Guiana Highlands.

▲ **Anteater**
There are three South American anteater species. All have long, sticky tongues, for eating ants and termites.

▲ **Big river**
The Amazon is the world's second-longest river, after the Nile, with a total length of over 6,400 kilometres.

Across the Brazilian border, these descend to the great basin of the Amazon River. This mighty river is fed by thousands of waterways which seep through the tropical growth of the world's largest rainforest, covering an area of over 330 million hectares. Broad and muddy, the Amazon flows eastwards to the Atlantic Ocean. Southern Brazil rises to the Brazilian Highlands and the tropical plateau of the Matto Grosso. This is drained in the south by the river Paraná.

◄ **Palms in the Pantanal**
In southwestern Brazil, the Pantanal wetlands are created by the seasonal flooding of the Paraguay River. The waters cover about 101,000 square kilometres of the Matto Grosso.

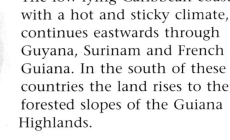

FACTS

BRAZIL
República Federativa do Brasil
Area: 8,547,403 sq km
Population: 161.2 million
Capital: Brasília
Other cities: São Paulo, Rio de Janeiro, Salvador, Belo Horizonte, Pôrto Alegre
Highest point: Neblina (3,014 m)
Official language: Portuguese
Currency: cruzeiro real

GUYANA
Co-operative Republic of Guyana
Area: 214,969 sq km
Population: 0.8 million
Capital: Georgetown
Other cities: New Amsterdam, Linden
Highest point: Roraima (2,772 m)
Official language: English
Currency: Guyana dollar

SURINAM
Republic of Suriname
Area: 163,265 sq km
Population: 0.4 million
Capital: Paramaribo
Other cities: Groningen, Nieuw Amsterdam
Highest point: Juliana Top (1,230 m)
Official language: Dutch
Currency: Surinam guilder

◄ Iguaçu Falls
The Iguaçu River plunges over hundreds of waterfalls and rocks before it joins the Paraná on the Brazil-Argentina border.

▲ Devil's Island
Just off the coast of French Guiana, this popular tourist resort was once a French prison colony.

▼ Flesh-eaters
Shoals of hungry piranha are found in the Amazon basin.

Map labels:

Gulf of Venezuela
Netherlands Antilles
Maracaibo
Lake Maracaibo
Caracas
Barcelona
Port of Spain
TRINIDAD & TOBAGO
GUYANA
ANDES MTS.
LLANOS
Orinoco
Pico Bolívar 5,002 m
VENEZUELA
Orinoco Delta
Angel Falls
Georgetown
GUYANA
Paramaribo
SURINAM
SURINAM
COLOMBIA
GUIANA HIGHLANDS
Cayenne
FRENCH GUIANA
FRENCH GUIANA
VENEZUELA
Orinoco
Branco
Pico da Neblina 3014 m
Negro
Japurá
Macapá
Marajó Bay
Marajó I.
Belém
São Marcos Bay
São Luís
Manaus
Santarém
Amazon
Tocantins
Teresina
Fortaleza
SELVAS
Madeira
Tapajós
Xingu
Natal
Recife
Juruá
Purus
Aripuanã
Araguaia
Parnaíba
SERTÃO
São Francisco
Jiparaná
Rio Branco
BRAZIL
Maceió
PERU
SERRA DOS PARECIS
Guaporé
Arinos
Sobradinho Reservoir
Salvador
BOLIVIA
MATO GROSSO PLATEAU
Cuiabá
Brasília
Goiânia
BRAZILIAN HIGHLANDS
BRAZIL
Uberlandia
Campo Grande
Paraná
Belo Horizonte
Campos
São Paulo
Santos
Rio de Janeiro
Cape Frio
PARAGUAY
Itaipu Res.
Itguaçu Falls
Curitiba
SERRA DO MAR
Florianópolis
ARGENTINA
Uruguay
Santa Maria
Pôrto Alegre
Patos Lagoon
URUGUAY
Mirim Lake

▲ Giant snake
The anaconda can grow to nine metres. It hunts in rivers and pools, strangling or drowning its prey.

▶ In eastern Brazil
This sandstone pillar at Vila Velha, in Brazil's Espirito Sánto state, has been weathered by wind and rain from the South Atlantic.

▶ Tapir
This rhinoceros-like mammal lives in the tropical rainforests.

135

Brazil and its neighbours

THE NORTHERN HALF OF SOUTH AMERICA IS DIVIDED UP INTO FOUR independent republics. These include Brazil, the largest country in the continent, Venezuela and two small Caribbean countries, Surinam and Guyana. Neighbouring French Guiana is still governed directly as a region of France.

This part of the world has huge cities and busy ports as well as remote forest and mountain areas that have never been properly explored. The region has great natural wealth in the form of Venezuelan oil, coal, iron ore, bauxite (for making aluminium), chrome, copper, gold and silver. Brazilian factories make cars and computers.

The rainforests of the Amazon provide timber, but rapid clearance of the forests by farmers, miners and loggers threatens to be a disaster not just for Brazil but for our planet as a whole. Beef cattle are raised on much of the land that has been cleared. Cattle are also reared on the Llanos grasslands of Venezuela.

▲ **Lake of oil**
The Venezuelan economy depends on Lake Maracaibo's oil reserves.

▲ **Pet parrot**
Many parrots are collected from the rainforest to be sold as pets.

▼ **A safe road**
Cattle are herded along the Transpantaneira, the only route through Brazil's Pantanal wetlands.

FACTS

VENEZUELA
República de Venezuela
Area: 912,050 sq km
Population: 22.3 million
Capital: Caracas
Other cities: Maracaibo, Valencia
Highest point: Pico Bolívar (5,002 m)
Official language: Spanish
Currency: bolívar

FRENCH GUIANA
Guyane Française
Overseas region of France
Area: 90,000 sq km
Population: 0.2 million
Capital: Cayenne
Official language: French
Currency: French franc

▼ **Space age**
The rocket launch site at Kourou in French Guiana is operated by the European Space Agency.

▼ **Paramaribo, Surinam**
Paramaribo, at the mouth of the Surinam River, is Surinam's capital and its biggest port. Nearly 50 percent of the population live there.

◄ Sugarcane
Workers harvest the sugarcane crop. Next, the stalks are crushed and soaked in water to produce a sugary liquid. When this is heated, it separates into brown crystals of cane sugar and sticky molasses.

▼ Cayenne pepper
This hot spice takes its name from the capital of French Guiana.

Brazil is the world's leading producer of coffee and sugarcane. Other crops include rice, maize, soya beans, cassava and citrus fruits. Cayenne, in French Guiana, is famous for its red hot pepper, while Guyana is known for the brown sugar first produced in the Demerara region.

Despite its resources, the region has faced huge economic problems. There has been severe inflation, rising debt and political corruption. Venezuela relies heavily on its oil, for which the price has fallen in recent years.

▼ Cassava plantation
The fleshy roots of the cassava plant are used to make flour; its leaves are eaten as a vegetable.

▼ Brazil nuts

Although some people are wealthy, many are desperately poor. In big cities, such as Caracas or Rio de Janeiro, families with little chance of finding work crowd into makeshift shanty towns and slums. Many of the indigenous peoples of remote areas, referred to as 'Indians', have been killed and had their land stolen. Their rivers have been poisoned and the forests where they hunt have been cut down.

► A vanishing world
The Amazon basin contains one-third of the world's surviving rainforest. This is one of the planet's most precious resources, but vast areas have already been destroyed.

Brazil and its neighbours

THE FIRST OR INDIGENOUS PEOPLES OF northern South America now make up only a very small portion of the population. It varies from five percent in Guyana to under two percent in Brazil. They mostly live in scattered communities, many in remote areas of rainforest where they hunt and fish. Groups include the Xingu, Yanomami, Shavanti and Kayapo, the coastal Caribs, Warrau and many others besides. Over 85 indigenous groups have been destroyed over the last 100 years.

Many people of the region are mestizos, of mixed European and indigenous descent. European settlement of the region has been from Portugal, Spain, Italy, the Netherlands, France, Britain and Germany. There are also sizeable black populations, of African descent. The Caribbean coast has a large Asian population, with many Javanese living in Surinam.

▲ **A rainforest village**
The round huts of the Yanomami may be seen in Venezuela and Brazil. Yanomami lands have been destroyed by mining since the 1970s.

▲ **Kayapo musician**
The Kayapo people live in northeastern Brazil, between the Araguaia and Xingu rivers.

▶ **Java comes to Surinam**
Many workers from Dutch colonies in Southeast Asia found work in Surinam after 1863. They brought their own culture, such as this ritual horse dance.

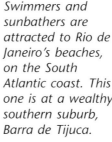

▲ **On the beach**
Swimmers and sunbathers are attracted to Rio de Janeiro's beaches, on the South Atlantic coast. This one is at a wealthy southern suburb, Barra de Tijuca.

◀ **River borders**
The Paraguay River flows past Corumba, a town on Brazil's frontier with Bolivia. The river goes on to form the Brazil-Paraguay, and Paraguay-Argentina borders.

▲ **Mixed descent**
Brazil is a multiracial society, with many people of mixed indigenous, European or African descent.

◄ Caracas
The Venezuelan capital, Caracas, has spread out from its old colonial centre to include city parks, high-rise offices and apartments, shopping malls and factories.

► Shanty town
São Paolo, home to over 15 million people, is Brazil's biggest city. Many desperately poor families live in makeshift housing and slums.

The region's official languages all date from its period of colonial rule – Portuguese in Brazil, Spanish in Venezuela, English in Guyana, Dutch in Surinam and French in French Guiana.

Brazil and Venezuela are largely Roman Catholic, but there are also Protestants and people of other faiths such as Hinduism and Islam. Spirit religions such as Candomblé, influenced by both African and Roman Catholic traditions, are popular in Brazil.

The fascinating mixture of cultures in the region has influenced festivals such as the five-day carnival in Rio de Janeiro, with its dazzling costume parades and dancing to Brazil's very own rhythm, the samba. Ethnic variety has also contributed to regional cooking, which combines African, Asian and European influences.

As in most Latin American regions, soccer is by far the most popular sport. It is practised on almost every beach, village clearing or street corner. Brazil has the most successful World Cup record of any nation.

▼ Carnival in Rio
For five days each year the Brazilian city of Rio de Janeiro is taken over by carnival, with costumed revellers dancing the samba in the streets.

▲ Football fans
Soccer is popular all over South America, and nowhere more so than Brazil, where the national team has won the World Cup four times.

Brazil and its neighbours

THE EARLY PEOPLES OF THE RAINFOREST LEFT BEHIND fewer reminders of their way of life than the peoples of the high Andes. They built with wood, creepers and straw instead of stone – materials which do not survive long in humid, tropical climates. Even so, archaeologists are only now beginning to realise the skill with which these peoples managed the forest, planting trees and seeds they could use for food, medicines and shelter.

▲ **Forest vines**
The forest provided building materials, foods and medicines to the first Amazonians.

The Spanish arrived in Venezuela in 1498, and the Portuguese in Brazil just two years later. Their arrival was a disaster for the indigenous peoples, many of whom died of European diseases. The French and Dutch occupied territory on the Caribbean coast during the 1600s and 1700s, with Britain capturing what is now Guyana in 1796. The Europeans planted sugar cane and other crops, and brought in slave labour from Africa to work on the plantations. Later, many Asian labourers were hired to work along the Caribbean coast.

▲ **Spanish Caracas**
Colonial buildings date back to the days when Venezuela was ruled by Spain.

▼ **Monte Serrat fort**
Several old forts may still be seen in Salvador, this Brazilian port was founded by the Portuguese in 1549.

▲ **Pedro Alvares Cabral**
This Portuguese navigator discovered Brazil by mistake, in 1500. His fleet, bound for India, was carried west by ocean currents.

▶ **Georgetown, Guyana**
The Caribbean port of Georgetown is Guyana's capital. City Hall, on the Avenue of the Republic, dates from the colonial period (1831–1970).

▼ The liberator

Simón Bolívar was born in Venezuela in 1783. In the 1820s, his armies helped to drive the Spanish out of South America.

◀ Opera House, Manaus

Manaus lies in the rainforest, on the Río Negro. Its ornate opera house dates from 1896, when the region was profiting from the rubber trade.

▼ Brasília

This purpose-built city, planned by architect Lucio Costa, became Brazil's capital in 1960. It was more centrally located than the previous capital, Río de Janeiro.

Venezuela broke away from Spanish rule between 1811 and 1821, after a war of independence led by freedom-fighter Simón Bolívar. Portugal became a kingdom in its own right in 1815 and was declared an independent empire seven years later. However the rich landowners threw out the royal family when it ordered an end to slavery. In 1889, Brazil became a republic. In the 1960s it moved its capital from Rio de Janeiro, on the Atlantic coast, to Brasília, a new, specially-built capital in the centre of the country on the Brazilian Highlands.

In 1946 French Guiana, until then chiefly famous for harsh prison settlements such as Devil's Island, became ruled on an equal status with mainland France. Guyana gained independence from Britain in 1970 and Surinam from the Netherlands in 1975.

▶ The oil president

Carlos Andréas Perez was president of Venezuela 1974–9, during the country's oil boom.

▼ Landless protest

Despite Brazil's vast natural resources, many millions of its citizens live below the poverty line. Here, landless farm-labourers join a protest rally in São Paulo.

AD	**TIMELINE**
400	*Marajoara culture, mouth of Amazon*
1498	*Spanish land in Venezuela*
1500	*Portuguese claim Brazil*
1530	*Portuguese colonize Brazil*
1593	*Spanish claim Surinam*
1600s	*Sugar plantations in Brazil: slave labour imported from Africa*
1602	*Dutch settle Surinam*
1604	*French settle Cayenne*
1620s	*Dutch settle Guyana*
1667	*Surinam becomes a Dutch colony*
1749	*Venezuelan revolt against Spain*
1815	*Brazil united with Portuguese kingdom*
1821	*Venezuela joins independent Colombia*
1822	*Brazil declares independence under Spanish prince, Pedro*
1825	*Pedro I recognised as emperor of Brazil*
1829	*Venezuela becomes independent state*
1831	*Guyana becomes colony of British Guiana*
1834	*Slavery abolished in British Guiana*
1888	*Slavery abolished in Brazil*
1889	*Brazil becomes a republic*
1910	*Oil discovered in Venezuela*
1946	*French Guiana becomes overseas department of France*
1960	*Brasília becomes capital of Brazil*
1966	*Guyana becomes independent*
1975	*Surinam becomes independent*
1992	*First Earth Summit (UN world environment conference), Río de Janeiro*

Argentina
and its neighbours

▲ **Elephant seals**

Paraguay lies at the heart of South America. It is crossed by the Paraguay River, which flows into the Paraná. In the west is the tropical wilderness of the Gran Chaco, and in the east both forest and grasslands, taken up by farms and ranches. Uruguay lies on the mouth of the river Plate, on the South Atlantic coast. It is a land of fertile plains and low hills.

In the northeast of Argentina is the rich farmland of the Mesopotamia region, lying between the Paraná and Uruguay rivers. The central north takes in part of the Gran Chaco, while the central eastern region is made up of open, rolling grassland. Known as the pampa, it is now used for ranching and agriculture. To the south again are the bleak, grassy plateaus and valleys of Patagonia.

The southern Andes form the border between Argentina and Chile, rising to the highest peaks in all the Americas. Chile lies between the western slopes and the South Pacific Ocean. This narrow coastal strip includes cool green regions, warm lands with a mild, Mediterranean climate, and the Atacama Desert, one of the driest spots on Earth.

Across the Strait of Magellan, the continent breaks up into the islands of Tierra del Fuego and the cold, grey seas off Cape Horn.

▲ **In Patagonia**
South of the Río Negro, Argentina is a land of wind-swept plateaus and valleys.

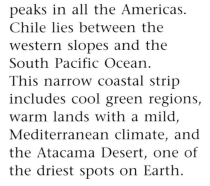

▲ **Atacama**
This coastal region of northern Chile is dry, barren desert. However it does contain valuable mineral resources.

▶ **Tierra del Fuego**
The southern tip of South America is a bleak peninsula occupied by both Argentina and Chile. The southernmost town in the world, Ushuaia, lies on the Beagle Channel.

142

FACTS

ARGENTINA
República Argentina
Area: 2,780,400 sq km
Population: 35.2 million
Capital: Buenos Aires
Other cities: Córdoba, La Plata, San Miguel de Tucumán
Highest point: Aconcagua (6,959 m)
Official language: Spanish
Currency: Argentinean peso

CHILE
República de Chile
Area: 756,626 sq km
Population: 14.4 million
Capital: Santiago
Other cities: Vina del Mar, Valparaíso, Concepción
Highest point: Ojos del Salado (6,880 m)
Official language: Spanish
Currency: Chilean peso

PARAGUAY
República del Paraguay
Area: 406,752 sq km
Population: 5.0 million
Capital: Asunción
Other cities: Ciudad del Este, Pedro Juan Caballero
Highest point: Villarrica (680 m)
Official language: Spanish
Currency: guaraní

URUGUAY
República Oriental del Uruguay
Area: 177,414 sq km
Population: 3.2 million
Capital: Montevideo
Other cities: Salto, Rivera
Highest point: Mirador Nacional (501 m)
Official language: Spanish
Currency: Uruguayan peso

FALKLAND ISLANDS
Islas Malvinas
British Overseas Colony
Area: 12,173 sq km
Population: 2,000
Capital: Stanley
Official language: English
Currency: Falkland Islands pound

In the southern Andes
The heavy ice of the Moreno Glacier grinds its way down to Lago Argentina, one of the most awesome sights in South America.

Canyon lands
Rivers have carved out canyons from the rocks of Patagonia. The Andes Mountains shield this vast, dry region of Argentina from the Pacific Ocean's rain-bearing winds.

Guano rocks
This rock, near Constitución in central Chile, is covered in guanos (the droppings of seabirds). In parts of South America, guano is a valuable resource. It is collected to make fertilizers.

CHILE

PARAGUAY

Prickly pear

URUGUAY

Armadillo

Curtain of spray
Argentina borders Brazil along the Iguazú (Iguaçu) River. A long stretch of the river is broken by islands, rapids and waterfalls.

SOUTH GEORGIA (U.K.)

ARGENTINA

N
W E
S

Volcano
Volcanic peaks rise above Lake Chungara, in Chile.

Brown pelican

143

Map labels

Arica
Iquique
ATACAMA DESERT
Calama
Antofagasta
BOLIVIA
Salta
Ojos del Salado 6,880 m
San Miguel de Tucumán
Copiapó
Catamarca
Santiago del Estero
La Rioja
Coquimbo
Pta. Lengua de Vaca
San Juan
Aconcagua 6,959 m
Mendoza
Valparaíso
Santiago
San Rafael
Rancagua
San Luis
Talca
Chillán
Concepción
Pta. Lavapié
CHILE
Neuquén
Temuco
Negro
Valdivia
Pta. de la Galera
Limay
Osorno
Puerto Montt
Chiloé I.
Chubut
C. Quilán
Chico
LOS CHONOS ARCHIPELAGO
Lake Buenos Aires
Deseado
Penas Gulf
Chico
PACIFIC OCEAN
Wellington I.
Santa Cruz
REINA ADELAIDA ARCHIPELAGO
Punta Arenas
Tierra del Fuego
Santa Inés I.
Ushuaia
Cape Horn

GRAN CHACO
Verde
Pilcomayo
Bermejo
PARAGUAY
Concepción
BRAZIL
Asunción
Cuidad del Este
Formosa
Paraguay
Resistencia
Alto Paraná
Corrientes
Posadas
MESOPOTAMIA
Salado
Mar Chiquito
Paraná
Córdoba
Concordia
Salto
Santa Fe
Paraná
Paysandú
Rosario
Uruguay
Río Cuarto
Negro
Buenos Aires
La Plata
Montevideo
URUGUAY
Río de La Plata
Pta. Norte
Cape San Antonio
PAMPAS
Mar del Plata
Cape Corrientes
ARGENTINA
Bahía Blanca
Bahía Blanca
Colorado
Viedma
San Matías Gulf
Valdés Peninsula
Rawson
Comodoro Rivadavia
San Jorge Gulf
C. Tres Puntas
Puerto Deseado
Puerto Santa Cruz
FALKLAND/MALVINAS ISLANDS
Bahía Grande
West Falkland
Stanley
East Falkland
Río Gallegos
Strait of Magellan
C. San Diego
SIERRA DE CÓRDOBA
ANDES MOUNTAINS
PATAGONIA

Argentina and its neighbours

THE SOUTHERN MAINLAND OF SOUTH AMERICA is divided into four independent republics – the small northern nations of Uruguay and Paraguay, the wide open spaces of Argentina and the long, narrow territory of Chile. The region has experienced long years of rule by brutal military regimes and there have also been wars over borders and territory. In 1982 Argentina went to war with the United Kingdom over ownership of the Falkland or Malvinas Islands, a small British colony in the South Atlantic Ocean. Today the region as a whole has moved to democratic systems of government, but faces many economic and political problems.

Paraguay raises cattle, and grows cotton. The leaves of a plant called the Paraguay holly are used to make a bitter tea called yerba maté, which is also very popular in Argentina. Uruguay, too, is cattle country, with fruit grown on the coastal plains.

▲ **Military rule**
The army held power in Argentina from 1976 to 1983, in Chile from 1973 to 1989, in Paraguay from 1954 to 1993, and in Uruguay from 1973 to 1984.

▲ **Atlantic outpost**
Sheep are raised on the remote British colony of the Falkland Islands, or Islas Malvinas.

▶ **Cattle Market**
A Chilean cowboy is called a huaso. Cattle are raised in the the central Chilean regions to the south of Santiago.

▲ **Salt lake**
The bone-dry air of Chile's Atacama Desert evaporates pools of water, leaving behind salt pans.

▶ **Valdés wildlife**
Magellan penguins may be seen off Argentina's Valdés peninsula, along with elephant seals and whales.

◀ **Yoked oxon**
Oxen are yoked in the traditional way on this farm near Coyhaique, in Patagonia.

▶ **City memorial**
This obelisk rises above the Plaza de la República in central Buenos Aires. It commemorates the founding of the city in 1536.

Itaipú dam
This is the world's biggest hydroelectric scheme, jointly operated by Argentina and Paraguay. Building the dam created a 1,350 sq km reservoir, at a cost of US$25 billion.

Argentina first became wealthy by exporting beef. Great herds of cattle are still raised on the pampa, and Patagonia raises sheep. The foothills of the Andes provide shelter for vineyards in the sunny region around Mendoza. Fisheries, oil and natural gas are important industries of the far south – Ushuaia is the most southerly town in the world. Industry is centred around the capital of Buenos Aires and is based on food processing, leather goods, textiles, chemicals and car manufacture.

Argentinian wines
The sheltered region around Mendoza has a mild, dry climate. Rivers and irrigation schemes have allowed vines to be grown here since the 1600s.

Chile's industry is based around the port of Valparaiso, where iron, steel, chemicals and textiles are produced, and the capital city of Santiago. The country has large reserves of copper in the north. The warm regions of central Chile produce wine and citrus fruits for export and the long coastline provides harbours for large fishing fleets.

Plaza del Congreso
At the heart of Buenos Aires, capital of Argentina, is the Palace of Congress. Congress is made up of two houses, the Senate and the Chamber of Deputies.

Santiago
Nine out of ten Chileans live in towns or cities. The capital is the country's largest city. It is home to nearly five million people.

Juicy fruit
The orchards of central Chile produce fruit for export, including oranges, lemons, apples, grapes and pears.

Copper kings
Chile is the world's biggest copper producer. The monster Chuquicamata open-cast mine is sited in the far north, in the Atacama Desert.

145

Argentina and its neighbours

NATIVE AMERICAN PEOPLES MIGRATED SOUTHWARDS through the continent in prehistoric times. They had reached central South America by 14,000BC and were in Patagonia by 12,000BC. While northern Chile came under the influence of the great civilizations of the northern Andes, the peoples of the far south lived by hunting on the grasslands or fishing the cold southern waters in their canoes.

The Spanish invaded the region in the 1500s. They were fiercely resisted by indigenous groups such as the Querandí of the pampa and the Mapuche of Chile. The region broke away from Spanish rule in the 1800s, but generals and warlords fought out regional and civil wars. There was also genocide, the organized murder of whole native peoples. The pampa of Argentina were taken over by huge ranches. They were worked by gauchos, wild-living cowboys of mixed Spanish-indigenous descent.

Immigrants poured into the region in the 1800s and 1900s – Spanish, Italians, Basques, Welsh, English, Germans, Swiss, Poles, Russians, Jews, Syrians. More recently there were arrivals from Japan, Korea and from other parts of South America.

▲ **Araucarian people**
The indigenous Araucanians of Chile include the Mapuche, who resisted conquest for over 300 years.

▼ **The gauchos**
The tough, bragging cowboys of the Argentinean pampas were called gauchos. Today's cowboys still honour the gaucho tradition.

▲ **Desert mummy**
Dead bodies from early South American civilizations have been preserved as mummies. This woman was found in the Atacama Desert.

▼ **La Boca**
This district of Buenos Aires, along the polluted Riachuelo waterway, was founded by poor Basque and Italian immigrants. Today it is a centre for artists, and the buildings are painted in bold, bright colours.

▲ **Andean church**
A white church stands on the edge of the Andes at Tocanao, in Chile's Antofagasta region. It is built from volcanic stone.

▼ Polo champions
Eduardo Heguy comes from a family of polo stars. Argentina has achieved international success in this exclusive sport.

▶ Armadillo lute
The shell of the armadillo is used to make a traditional stringed instrument.

▶ Eva Perón (1919–52)
This former actress married Argentinean politician Juan Perón in 1945. 'Evita' was a popular and powerful influence on the country.

Today the descendants of all these peoples dominate the region. Indigenous peoples now make up a tiny percentage of the population, but their cultures survive in some areas. The native Guaraní language is widely spoken in Paraguay. Spanish is spoken everywhere, but other European languages, from Italian to Welsh, may also be heard. Far out in the South Atlantic, the Falkland Islanders remain English in their speech and customs. The mainland nations are mostly Roman Catholic, with Protestant and Jewish minorities.

The region has produced sporting champions in motor racing, soccer and tennis, and the Andes Mountains are increasingly popular for skiing and outdoor pursuits. Both Chile and Argentina have produced great writers and poets in the last century, including the Argentinian Jorge Luis Borges (1899–1986) and the Chilean novelist Isabel Allende (*b*.1942).

▲ Viña del Mar
South American tourists flock to this resort, near Valparaíso in Chile, for the beaches and casino.

▶ Grand Prix
Motor racing has been a passion since the 1950s, when Argentinian Juan Fangio was five times Grand Prix champion.

AD	**TIMELINE**
1516	*Río de la Plata discovered by the Spanish*
1536	*First founding of Buenos Aires by Spanish*
1537	*Spanish found Asunción*
1541	*Querandí people attack Buenos Aires Spanish conquer Chile*
1543	*Rebuilding Buenos Aires (until 1580)*
1726	*Spanish build fort at Montevideo*
1811	*Paraguay becomes independent*
1816	*Argentina becomes independent*
1818	*Chile becomes independent*
1825	*Uruguay declares independence (recognized 1828)*
1857	*European settlement of the pampa*
1865	*Paraguay fights Uruguay, Argentina and Brazil (until 1870)*
1878	*Wars against Native Americans in Argentina (until 1883)*
1879	*War of the Pacific (until 1884): Chile defeats Peru and Bolivia*
1946	*Juan Perón becomes Argentinean president*
1954	*Military coup in Paraguay*
1973	*Allende overthrown by General Augusto Pinochet in Chile*
1976	*Military rule in Argentina: imprisonment and murder of opponents*
1982	*Argentina fights United Kingdom over the Falkland (Malvinas) Islands*
1983	*Democracy restored in Argentina*
1992	*Democracy restored in Paraguay*

ASIA

Asia, the world's largest continent, covers about 30 percent of the world's land area. It is bordered by Europe and Africa in the west and by the Pacific Ocean in the east. There are chains of tiny volcanic islands in the east, an area known as 'The Ring of Fire' because it is so prone to earthquakes.

Asia's northwestern limits are marked by the Ural and Caucasus mountains. Lands in the far north overlap the Arctic Circle and much of the region is tundra, frozen solid for most of the year. Farther south lies a swathe of evergreen forest and, beyond that, the steppes – open, fertile grasslands. Little rain reaches Central Asia, so much of it is desert.

The world's highest mountain ranges, the Himalayas and the Karakorams, form a snowy barrier to the north of the Indian peninsula, and include the world's highest peak, Mount Everest. To the south, the world's greatest rivers, the Ganges and Brahmaputra, run towards the warm Indian Ocean. They often flood, depositing rich, alluvial soil over the wide delta.

Asia's southwest coastline is cooled by breezes from the Red Sea and the Mediterranean, but the lands and islands of southern Asia have a tropical climate. There, the winters are hot and dry, while in summer stormy winds called monsoons bring heavy rains.

Asia's many ethnic groups include the Arabs of Southwest Asia, the Hindu people of India and the Chinese in Eastern Asia. The continent was the home of major early civilizations and the birthplace of all the major world religions.

Many Asians are poor and about 60 percent live by farming. But in the last 50 years, some countries have developed quickly and raised their living standards. In parts of Southwest Asia, wealth has come from oil, while countries in Eastern and Southeast Asia have developed industries. Japan is Asia's most industrialized country, but some other countries in Eastern Asia have also developed quickly. Eastern Asia will dominate the world economy in the 21st century.

▲ **Space Monument, Russia**

▲ **Puffer fish**

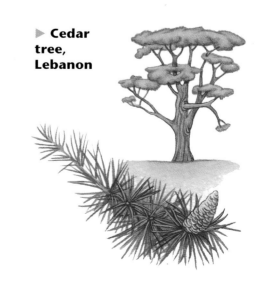

▶ **Cedar tree, Lebanon**

▶ **Urgup Cones, Turkey**

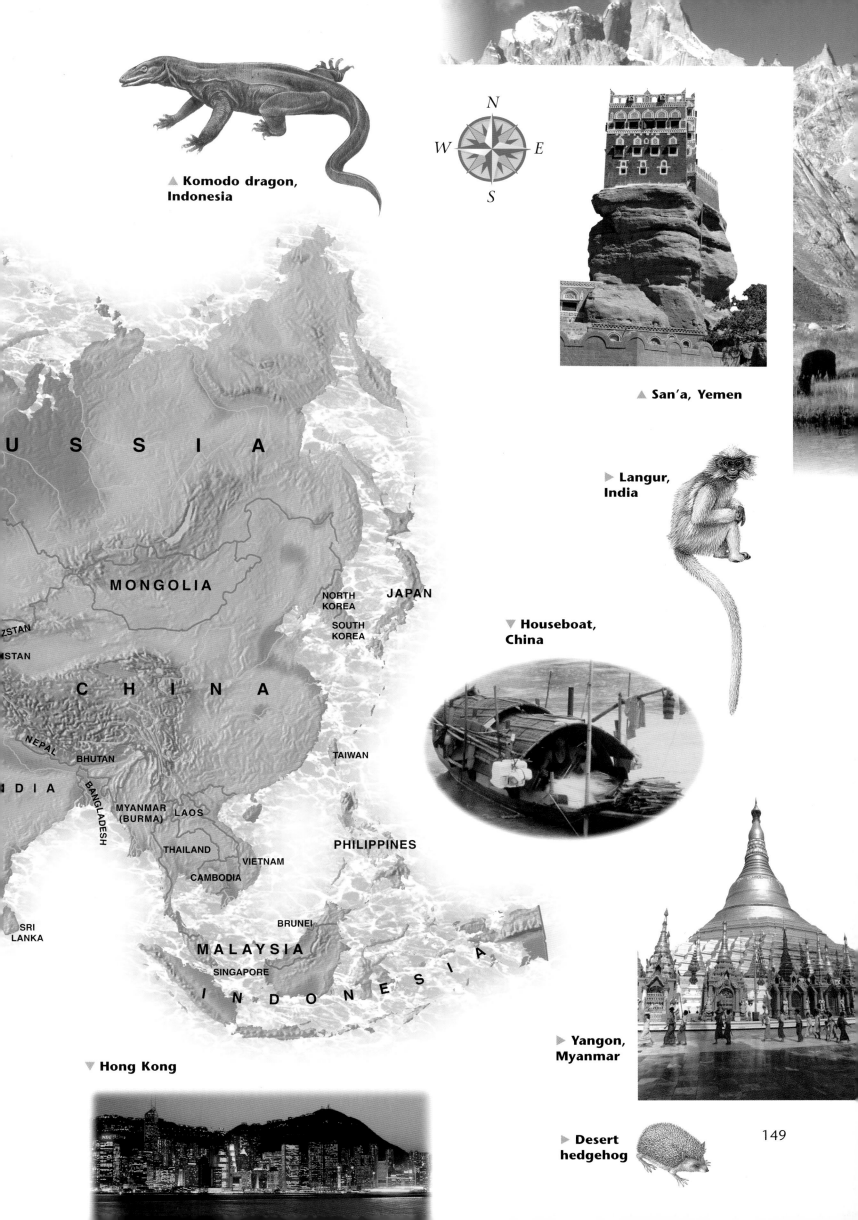

▲ Komodo dragon,
Indonesia

N
W *E*
S

▲ San'a, Yemen

► Langur,
India

RUSSIA

MONGOLIA

NORTH KOREA
SOUTH KOREA
JAPAN

▼ Houseboat,
China

ZSTAN
STAN

CHINA

NEPAL
BHUTAN
DIA
BANGLADESH
MYANMAR (BURMA)
LAOS
THAILAND
VIETNAM
CAMBODIA
PHILIPPINES

TAIWAN

SRI LANKA

BRUNEI

MALAYSIA
SINGAPORE

INDONESIA

► Yangon,
Myanmar

▼ Hong Kong

► Desert
hedgehog

149

Russia and its neighbours

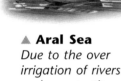

▲ **Grey wolf**

Eastern Europe runs into Asia without any obvious borders, apart from mountains and rivers. The Russian Federation and some of its neighbours straddle both continents.

Belarus lies on the great plain that stretches from Germany to Russia. To the south lie the steppes of Ukraine and Moldova, natural grasslands now harvested for wheat. Ukraine borders the Carpathian mountains in the west and the warm Black Sea in the south. The Caucasus Mountains run southeastwards through Georgia, Armenia and Azerbaijan towards the Caspian, the world's broadest inland sea. To the east of the Caspian Sea lie the Central Asian republics of Kazakhstan, Turkmenistan, Uzbekistan and Kyrgyzstan. Here, dusty grasslands give way to deserts and soaring peaks.

▲ **Aral Sea**
Due to the over irrigation of rivers and evaporation the Aral Sea is shrinking.

The Russian Federation is the biggest country in the world, stretching far beyond the Ural Mountains. It includes the whole of northern Asia, being bordered by the Pacific Ocean in the east. The northern coast is mostly locked in Arctic ice. Treeless plain, or tundra, is taken over by a great belt of forest, or taiga, made up of spruce and birch trees. In the southwest there are steppes and in the southeast, deserts and mountains. Russia is drained by great rivers such as the Don, the Volga, the Ob, Irtysh and Lena.

◀ **Ancient ruins**
The kingdom of Sogdiana was conquered by Alexander the Great in 328BC. Remains found here, at Pendzhikent in Uzbekistan, show that this was a crossroads of trade between India, China and Persia (modern Iran).

▶ **Plant life**
Polar plants are low-lying, to stay out of the bitter Arctic winds.

FACTS

ARMENIA
Haikakan Hanrapetoutioun
Area: *29,800 sq km*
Population: *3.8 million*
Capital: *Yerevan*
Other cities: *Gyumri, Ejmiadzin*
Highest point: *Aragats (4,090 m)*
Official language: *Armenian*
Currency: *Dram*

AZERBAIJAN
Azerbaijchan Respublikasy
Area: *86,600 sq km*
Population: *7.6 million*
Capital: *Baku*
Other cities: *Ganka, Sumgait*
Highest point: *Bazar Dyuzi (4,466 m)*
Official language: *Azeri*
Currency: *Manat*

BELARUS
Respublika Belarus
Area: *207,600 sq km*
Population: *10.3 million*
Capital: *Minsk*
Other cities: *Gomel, Pinsk*
Official language: *Belarussian*
Currency: *Rouble*

GEORGIA
Sakartvelos Respublika
Area: *69,700 sq km*
Population: *5.4 million*
Capital: *Tbilisi*
Other cities: *Kutaisi, Rustavi*
Highest point: *Shkhara (5,201 m)*
Official language: *Georgian*
Currency: *Lari*

KAZAKHSTAN
Qazaqstan Respublikasï
Area: *2,717,300 sq km*
Population: *16.5 million*
Capital: *Almaty*
Other cities: *Qaraghandy, Shymkent*
Highest point: *Tengri (6,398 m)*
Official language: *Kazakh*
Currency: *Tenge*

▲ **Ananuri Castle, Georgia**
Georgia became a powerful kingdom in the early Middle Ages. It was absorbed into the Russian empire in the 1800s.

▲ **Ukrainian sunshine**
River shores on the Dnepr and Dnestr and warm coastlines on the Black Sea attract Eastern European holidaymakers to Ukraine.

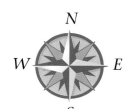

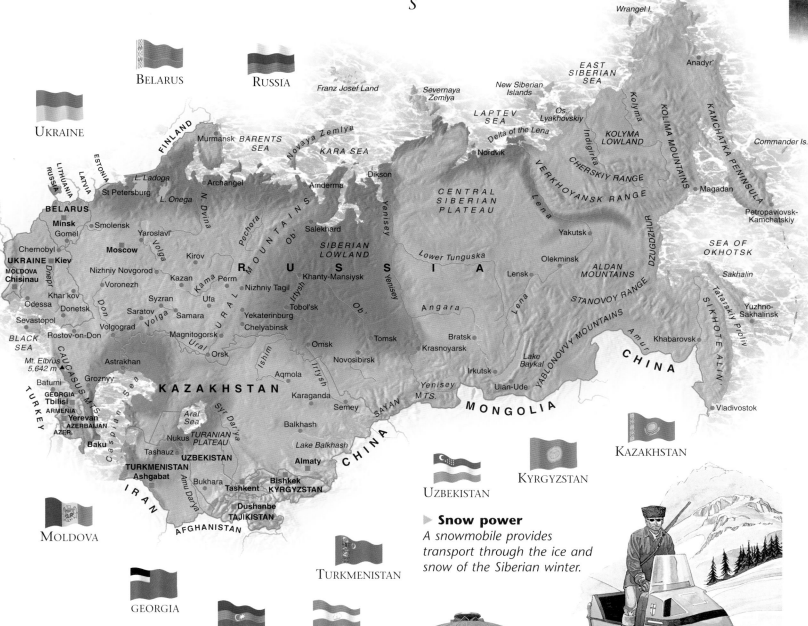

BELARUS

RUSSIA

UKRAINE

Wrangel I.

Franz Josef Land

Severnaya Zemlya

New Siberian Islands

EAST SIBERIAN SEA

Anadyr'

LAPTEV SEA

Os. Lyakhovskiy

Commander Is.

FINLAND

Murmansk

BARENTS SEA

Novaya Zemlya

KARA SEA

Delta of the Lena

Nordvik

Dikson

KOLYMA LOWLAND

Kolyma

KOLIMA MOUNTAINS

KAMCHATKA PENINSULA

LITHUANIA
RUSSIA
LATVIA
ESTONIA

L. Ladoga

St Petersburg

L. Onega

Archangel

Amderma

Salekhard

CENTRAL SIBERIAN PLATEAU

Lena

INDIGIRKA

CHERSKIY RANGE

VERKHOYANSK RANGE

Magadan

BELARUS

Minsk

Gomel

Smolensk

Yaroslavl'

N. Dvina

Pechora

URAL MOUNTAINS

Ob'

SIBERIAN LOWLAND

Yenisey

Yakutsk

Olekminsk

Petropavlovsk-Kamchatskiy

Chernobyl

Moscow

Kirov

Volga

Nizhniy Novgorod

SEA OF OKHOTSK

UKRAINE

Kiev

MOLDOVA

Chisinau

Voronezh

Kazan

Kama

Perm

Nizhniy Tagil

Khanty-Mansiysk

Irtysh

Yekaterinburg

Tobol'sk

Ob'

Lower Tunguska

Lensk

ALDAN MOUNTAINS

DZUGDZHUR

Sakhalin

Khar'kov

Syzran

Ufa

R U S S I A

Yenisey

Angara

Bratsk

STANOVOY RANGE

Khabarovsk

SIKHOTE ALIN

Odessa

Donetsk

Saratov

Samara

Chelyabinsk

Tomsk

Krasnoyarsk

Yuzhno-Sakhalinsk

Tatarskiy Proliv

Sevastopol

Volgograd

Magnitogorsk

Omsk

Don

Volga

Ural

Orsk

Novosibirsk

Irkutsk

Ulan-Ude

YABLONOVVY MOUNTAINS

Amur

Vladivostok

BLACK SEA

Rostov-on-Don

Astrakhan

Ishim

Irtysh

Aqmola

Yenisey MTS.

Lake Baykal

CHINA

Mt. Elbrus 5,642 m

Groznyy

CAUCASUS MTS

K A Z A K H S T A N

Karaganda

SAYAN MTS.

MONGOLIA

Batumi

Caspian Sea

GEORGIA

Tbilisi

ARMENIA

Yerevan

AZERBAIJAN

AZER.

Baku

Semey

Aral Sea

Nukus

Syr Dar'ya

TURANIAN PLATEAU

Balkhash

Lake Balkhash

CHINA

TURKEY

Tashauz

UZBEKISTAN

Almaty

KAZAKHSTAN

IRAN

TURKMENISTAN

Ashgabat

Bukhara

Amu Dar'ya

Tashkent

Bishkek

KYRGYZSTAN

UZBEKISTAN

KYRGYZSTAN

Dushanbe

TAJIKISTAN

AFGHANISTAN

MOLDOVA

▶ **Snow power**
A snowmobile provides transport through the ice and snow of the Siberian winter.

TURKMENISTAN

GEORGIA

AZERBAIJAN

TAJIKISTAN

ARMENIA

▶ **Tashkent market**
The Uzbekistan capital is an ancient centre of trade and of textile manufacture.

151

Russia and its neighbours

UNTIL 1991 ALL THE COUNTRIES listed in this section were part of a single country, the Soviet Union. This was formed in 1922, after a revolution in 1917 had brought communists to power in Russia. Communist rule ended in 1991 and the country was renamed the Russian Federation.

▲ The big harvest
The grasslands stretching from Ukraine through southern Russia are called steppes. Wheat is now grown in their rich black soil.

At that time, many of the border regions broke away to become independent nations. However Russian armed forces prevented some other bids for independence, as in the Chechenya region. A Commonwealth of Independent States (CIS) was formed to keep up economic links between the new nations.

▲ Snowy winters
Most of the Russian Federation experiences long winters with heavy snowfall. Average January temperatures vary from –9°C in Moscow to –14°C in Vladivostok.

The region as a whole is immensely rich in resources, including timber, oil, natural gas, gold and diamonds. Transport is a major problem in such a vast country, where many resources lie in remote areas of wilderness, such as the Asian region of Siberia. The Trans-Siberian rail network links Moscow, the Russian capital, with the Pacific port of Vladivostok.

▶ The Summer Palace
The Summer Palace, built on one of the islands that make up the great city of St Petersburg (Leningrad), was built for Tsar Peter the Great between 1710 and 1714.

▼ Nuclear disaster
A nuclear power plant at Chernobyl, in Ukraine (then part of the Soviet Union) melted down in 1986. Deadly radiation was blown across northern Europe.

▶ Trans-Siberian Express
The Trans-Siberian Railway runs between Moscow and Vladivostok. It was opened in 1905 and played an important part in Russian history. It remains the world's longest track, at 9,438 kilometres. The whole route takes eight days to travel. It now links up with services to Mongolia and China.

▼ The Winter Palace
The St Petersburg residence of the tsars was built in 1762 and restored in 1839. With the adjoining Hermitage, it now houses one of the world's great art collections.

Uzbekistan is the world's fourth-largest producer of cotton. Here, the crop is harvested and processed at Urgench, in the west of the country.

Cereal crops grow well in the fertile west and south, while warm southern lands such as Georgia produce fruits and wine. The Central Asian lands grow cotton. Industries are mostly based in the Eastern European part of the region and include iron and steel, truck manufacture, shipbuilding and mining.

During the years of communist rule, the Soviet Union was rapidly transformed from a largely undeveloped country into a great industrial power. However by the 1980s, Russian industry was old-fashioned, inefficient and polluting the environment. Little was done to solve these problems in the 1990s. Changes were too rapid, unemployment was high and many businesses were controlled by criminals or corrupt officials. Output fell disastrously and wages went unpaid.

Even so, Russia remains one of the most powerful countries in the world and the poor countries grouped around the Caspian and Aral Seas may soon be transformed by new oil wealth.

FACTS

MOLDOVA
Republica Moldova
Area: *33,700 sq km*
Population: *4.3 million*
Capital: *Chisinau*
Other cities: *Tiraspol, Beltsy*
Highest point: *Balaneshty (429 m)*
Official language: *Moldovan*
Currency: *Leu*

RUSSIAN FEDERATION
Rossiskaya Federatsiya
Area: *17,075,400 sq km*
Population: *147.7 million*
Capital: *Moscow*
Other cities: *St Petersburg, Novosibirsk*
Highest point: *Elbrus (5,642 m)*
Official language: *Russian*
Currency: *Rouble*

TAJIKISTAN
Repoblika i Tojikiston
Area: *143,100 sq km*
Population: *5.9 million*
Capital: *Dushanbe*
Other cities: *Kulyab, Khodzhent*
Highest point: *Mt. Kommunizma (7,495 m)*
Official language: *Tajik*
Currency: *Rouble*

▲ **Shopping in Moscow**
In the days of the Soviet Union, grand-looking state-owned stores offered little choice for consumers. Economic problems continue today.

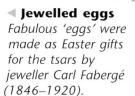

◄ **Jewelled eggs**
Fabulous 'eggs' were made as Easter gifts for the tsars by jeweller Carl Fabergé (1846–1920).

Russia and its neighbours

THE RUSSIANS ARE A SLAVIC PEOPLE. THEY MAKE UP 80 per cent or so of the population in the Russian Federation. The remaining 20 per cent is made up of over 150 minority groups. Some of these peoples, including the Chukchee, Evenks, Nenets and Saami, survive in the sparsely-populated Russian Arctic. Some of the Arctic peoples live by herding reindeer or by hunting. Amongst Russia's many other ethnic groups are the Cossacks of the steppes, the Kalmuk, Yakut, Tartars, Chechens, Cherkess, Dagestanis, Bashkir, Kets, Jews, Tuvinians and Buryats.

To the west of the new Russian borders are the Slavic Belarussians and Ukrainians and also the Moldovans, who are related to the Romanians. South, beyond the Caucasus, are the Georgians, Osset, Armenians and Azeris. To the east of the Caspian Sea are the Kazakhs, Turkmen, Uzbek, Kyrgyz and Tajik.

Despite the Russian Federation's wide-open spaces, most of its population lives to the west of the Urals, and three-quarters of them in cities and towns. The bitterly cold winters make life hard in many country regions.

▲ The Cossacks
A people of the steppes, the Cossacks were plunderers, famed for their riding skills and wild dances. They were later recruited by the tsars as fierce soldiers.

Peter Tchaikovsky

Leo Tolstoy

▲ The Arts
Russia has a rich history of writers, musicians and artists. Peter Ilyich Tchaikovsky (1840-93) composed memorable music and Leo Tolstoy (1828-1920) is considered to be one of the world's finest novelists.

◄ Arctic travel
Dogs, reindeer and horses were traditional providers of haulage in the Russian Arctic. Rivers may freeze so hard that trucks use them as highways.

▲ Ice hockey
Ice hockey is a popular sport in the Russian Federation, and national (and earlier USSR) teams are record-breaking winners of world-class and Olympic championships.

► Russian ballet
St Petersburg has a history of ballet dating back to 1738. Many great works were performed in the 1800s. Its most famous company is the Maryinsky (formerly Kirov) Ballet.

A warrior's tomb
The Gur-e Amir mausoleum in Samarkand, Uzbekistan, is decorated with mosaic tiles. It commemorates the Mongol emperor Timur the Lame or Tamerlane (1336–1405), who conquered most of Central Asia. He died marching eastwards to China.

Christianity is the main religion in the west and southwest of the region, with most people following Eastern Orthodox forms of worship. Onion-domed churches, some roofed with gold, may be seen in many towns. Other Christian groups also exist in the region. The Central Asian peoples are mostly Muslims and there are also Jews and Buddhists. Many religions were suppressed in the early days of the communist era.

▲ **Orthodox priests**
The Orthodox form of Christianity was brought to the Slavs by the Greek missionary St Cyril (AD827–69).

FACTS

TURKMENISTAN
Türkmenistan
Area: 488,100 sq km
Population: 4.6 million
Capital: Ashkhabad
Other cities: Charjew, Dashkhovuz
Highest point: Köpetdeg Range
Official language: Oguz Turkic
Currency: Manat

UKRAINE
Ukrayina
Area: 603,700 sq km
Population: 50.7 million
Capital: Kiev
Other cities: Kharkiv, Odessa
Highest point: Goverla (2,061 m)
Official language: Ukrainian
Currency: Karbovanets

UZBEKISTAN
Uzbekiston Respublikasi
Area: 447,400 sq km
Population: 23.2 million
Capital: Tashkent
Other cities: Samarkand, Bukhara
Highest point: Beshtor Peak (4,229 m)
Official language: Uzbek
Currency: Som

▲ **Russian dolls**
Hollow wooden dolls, each nesting inside another, are a Russian tradition and a popular souvenir of any visit to the Russian Federation.

During the Middle Ages, the Eastern Orthodox Church produced great works of art in the form of icons (holy pictures) and its worship is marked by fine choral singing. The arts have always been valued in Russia. St Petersburg and Moscow have been centres of ballet. Great Russian musical composers included Tchaikovsky (1840-93), while writers included the great Leo Tolstoy (1828-1910), author of *War and Peace* and *Anna Karenina*.

▶ **Musical tradition**
Traditional folk musicians perform in St Petersburg. The balalaika is an instrument with a triangular body and a neck like a guitar.

▶ **St Basil's Cathedral**
Dating back to 1555, the colourful onion-shaped domes of St Basil's rise above Red Square, in the centre of Moscow, the Russian capital.

Russia and its neighbours

THE STEPPES WERE SETTLED IN PREHISTORIC TIMES. About 2,500 years ago they were home to horseback warriors known as Scythians. In about AD400 Slavs began to settle western Russia. Swedish Vikings known as Rus traded along the Volga and Dniepr rivers and as far east as the Aral Sea. They gave their name to Russia.

The first Russian states were based on trading posts at Kiev (in today's Ukraine) and Novgorod. Moscow was founded in 1147. In the 1200s the Slavs came under fierce attack by Mongol armies from Asia.

Russia grew strong enough to resist the Mongols and build up its own empire. Its rulers were called tsars. In the 1600s Tsar Peter I, the 'Great', modernized Russia and built the fine city of St Petersburg. Russian armies were soon clashing with Turkey to the south. The empire spread eastwards into Siberia and westwards into Poland. It was invaded in turn by France in 1812.

▲ **Holy icons**
Painting an icon was considered to be an act of worship in itself.

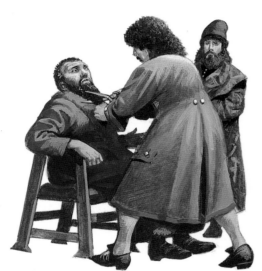

▲ **Peter the Great**
Born in 1672, Tsar Peter I travelled widely in Western Europe. By the time of his death in 1725, Russia had been transformed into a powerful, modern state.

▲ **Cutter of beards**
Peter I abolished the special powers of Russia's old aristocratic families. Their long beards, symbols of privilege, were cut off.

▼ **October Revolution**
Centuries of injustice under the rule of the tsars came to an end in 1917, when revolutionary workers, students and soldiers seized power in Russia.

◀ ▲ **Lenin and his tomb**
Vladimir Ilyich Ulyanov, known as Lenin, travelled from exile in Switzerland to lead the Bolsheviks in Russia's October Revolution. His tomb is in Moscow's Red Square.

▲ **Red Army leader**
Leon Trotsky (1879–1940) was a Marxist revolutionary who led the communist Red Army in Russia's civil war (1917–20). He was later exiled and murdered on the order of Joseph Stalin.

◄ In memory
Beneath the walls of the Kremlin, Moscow's medieval fortress, a flame burns in memory of the 'unknown soldier'. The Soviets suffered more casualties than any other nation in World War II.

▶ Space Age
A monument to space exploration stretches skywards. The Soviet Union was the first nation to launch a satellite into the Earth's orbit.

A hundred years ago, the Russian people had few political rights. Their ruling class resisted calls for change. Revolution broke out in 1905 and again in 1917, when the tsar and his family were shot dead. A socialist group called the Bolsheviks, led by Vladimir Ilyich Lenin (1870–1924), gained control of the whole country during a civil war which lasted until 1920.

Lenin died in 1924 and his successor was a ruthless politician called Joseph Stalin (1879–1953). Many Russians starved or died in labour camps under his rule. In 1941, Germany invaded. The Russian people resisted bravely and by 1945 the Germans had been pushed back into western Europe.

After the war, the Soviet Union was very powerful. It effectively controlled Eastern and Central Europe and its technology was advanced enough by 1961 to send the first person into space. The United States and its allies constantly challenged this power, during a 'Cold War', which lasted until the 1980s.

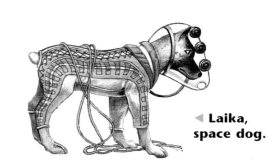

◄ Laika, space dog.

◄ Yeltsin years
In 1990 Boris Yeltsin (b.1931) became Russian president, presiding over the end of communism and break up of the old Soviet Union. As his health failed, the Russian Federation faced political unrest and increasing economic chaos.

▶ New conflicts
As new states were created from the ruins of the Soviet Union, there were violent clashes over territory and borders. Between 1989 and 1994, Armenia fought with Azerbaijan over the territory of Nagorno-Karabakh.

TIMELINE

AD

900s	Vikings found states in Russia and Ukraine
988	Orthodox Christianity introduced
1223	Mongols invade steppes
1462	Ivan the Great frees Russia from Mongols (until 1505)
1547	Ivan the Terrible becomes tsar
1881	Russian Tsar Alexander I assassinated
1905	Russia defeated by Japan; revolution in Russia, democratic reform
1917	October Revolution in Russia
1922	Soviet Union (USSR) founded
1924	Death of VI Lenin
1936	Joseph Stalin terrorizes opponents
1939	Soviet pact with Nazi Germany
1940	USSR annexes Baltic states
1941	USSR joins Allies in 'Grèat Patriotic War' against Nazi Germany (until 1945)
1946	USSR controls Eastern and Central Europe; Cold War with west (until 1989)
1979	USSR invades Afghanistan
1985	Mikhail Gorbachev introduces reforms
1989	Central Europe breaks away from USSR
1991	Break up of USSR; Russian Federation founded

Southwest Asia

◀ **Jerboa**

▲ **Cedar tree, Lebanon**

Southwest Asia, which is sometimes called the Middle East or the Near East, is a large region bordered by the Mediterranean and Black seas in the north, by the Red Sea in the southwest and by the Indian Ocean in the southeast. Thinly populated deserts and dry grassland make up much of the region. The most fertile regions are in the north. They include the Mediterranean island of Cyprus.

The northern part of Southwest Asia is mountainous, with high ranges extending through Turkey into Iran, where they include the Elburz Mountains in the north and the Zagros Mountains which run southeast through western Iran. The Elburz Mountains contain Mount Damavand, the region's highest peak. The Arabian peninsula also has mountain ranges, the longest of which borders the Red Sea. This range extends into southern Yemen. Inland from the mountains are large plateaux.

▲ **Wild cats**
Sand cats live throughout Southwest Asia and also Africa.

Most of Southwest Asia has a hot, desert climate. The dry Arabian peninsula contains one of the world's bleakest sandy deserts. It is called the Rub 'al Khali, or 'Empty Quarter'. Iran also has large deserts, some of which are covered by salt, which has been spread over the surface by occasional flood waters. Other dry areas, called steppe, are covered by grasses and low shrubs.

Turkey has a wetter climate than most of the region. Its coastline, like all the other lands bordering the Mediterranean Sea, has hot, dry summers and rainy, mild winters. But the dry, grassy interior plateaux of Turkey have bitterly cold winters, when snowstorms are common.

Water is scarce in much of Southwestern Asia, which has few major rivers. The two longest rivers are the Euphrates and Tigris. They rise in the mountains of Turkey and then flow across the dry country of Iraq. The valleys of these rivers were the sites of some major early civilizations.

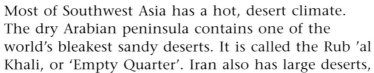

▶ **Citadel, Aleppo**
This citadel in the ancient city of Aleppo, northern Syria, was built in the Middle Ages.

FACTS

BAHRAIN
Dawlat al Bahrayn – State of Bahrain
Area: *694 sq km*
Population: *599,000*
Capital: *Al Manamah*
Other cities: *Muharraq*
Highest point: *Jabal ad Dukhan (135 m)*
Official language: *Arabic*
Currency: *Bahraini dinar*

CYPRUS
Kypriaki Dimokratia – Republic of Cyprus
Area: *9,251 sq km*
Population: *740,000*
Capital: *Nicosia*
Other cities: *Limassol, Larnaca*
Highest point: *Mount Olympus (1,952 m)*
Official languages: *Greek, Turkish*
Currency: *Cyprus pound*

IRAN
Jomhuri-e-Eslami-e-Iran – Islamic Republic of Iran
Area: *1,633,188 sq km*
Population: *62,509,000*
Capital: *Tehran*
Other cities: *Mashhad, Esfahan*
Highest point: *Mount Damavand (5,604 m)*
Official language: *Farsi (Persian)*
Currency: *Rial*

IRAQ
Joumhouriya al 'Iraqia – Republic of Iraq
Area: *438,317 sq km*
Population: *21,366,000*
Capital: *Baghdad*
Other cities: *Mosul, Kirkuk*
Highest point: *Haji Ebrahim (3,600 m)*
Official language: *Arabic*
Currency: *Iraqi dinar*

Urgup Cones
The landscape of the Goreme Valley, central Turkey, is dotted with strange, volcanic rock formations.

Esfahan, Iran
The city of Esfahan in Iran is famous for its beautiful mosques and other buildings.

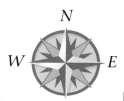
IRAQ

BLACK SEA

Istanbul
Gallipoli
Bursa
Sakarya
Samsun
PONTIC MOUNTAINS
Eskisehir
Ankara
Izmir
Tuz Lake
T U R K E Y
Kayseri
Konya
TAURUS MTS.
Antalya
Adana
Gaziantep
Diyarbakir
▲ Mt. Ararat 5,185 m
Aras

Lake Van

CASPIAN SEA

Tabriz
Rasht
Babol

TURKMENISTAN

IRAN

Nicosia
CYPRUS
Limassol
Tripoli
Aleppo
Mosul
S Y R I A
Euphrates
Tigris
As Sulaymaniyah
Kirkuk
Lake Urmia
ELBURZ MTS.
▲ Mt. Damavand 5,604 m
Mashhad

LEBANON
Beirut
Homs
Damascus
SYRIAN DESERT
Hamadan
Tehran
Qom
Dasht-e-Kavir

AFGHANISTAN

Haifa
ISRAEL
Tel Aviv
Jerusalem
Amman
I R A Q
Baghdad
Bakhtaran
Kashan
Esfahan
I R A N

JORDAN
Karbala
An Nasiriyah
Ahvaz
Yazd
Kerman
Dasht-e-Lut

EGYPT
Elat
Al Jawf
Sakakah
Basra
Abadan
KUWAIT
Kuwait
Shiraz
Bushehr
Zahedan

A N   N A F U D
Ad Dahna
Bandar Abbas

Buraydah
Shaqra
Ad Damman
Bandar e Lengeh
Jask
Strait of Hormuz

The Gulf
Al Manamah
BAHRAIN
QATAR
Doha
Dubai
Gulf of Oman

Medina
Riyadh
Abu Dhabi
UNITED ARAB EMIRATES
Muscat

HIJAZ
S A U D I   A R A B I A
▲ Jabal Ash Sham 3,035 m
Sur
O M A N

Jiddah
Mecca
ASIR
RED SEA
Tihamah
Masirah I.

▲ Jabal Sawda 3,133 m
*R u b '   a l   K h a l i
(E m p t y   Q u a r t e r)*
Salalah

Jaza'ir Farasan
Kuria Muria Is.

Tarim
Hadramaut
Al Hudaydah
San'a
Al Mukalla

Al Jawf
Bab al Mandab
Aden
Gulf of Aden
Y E M E N
Socotra (YEMEN)
'Abd al kuri

TURKEY

CYPRUS

LEBANON

ISRAEL

SYRIA

JORDAN

SAUDI ARABIA

IRAN

KUWAIT

BAHRAIN

OMAN

UNITED ARAB EMIRATES

QATAR

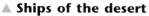

Ships of the desert
Camels can go long distances without water. They are used to transport goods throughout Southwest Asia.

Palace in Yemen
The Wadi Dahr Palace is perched on a steep rock in San'a, capital of Yemen.

YEMEN

Fort in Oman
Most Omani towns contain an old fort. Today, many of them are crumbling.

Desert hedgehog

159

Southwest Asia

SOUTHWEST ASIA CONTAINS 15 COUNTRIES. THE Arabs are the largest single group of people. Other large groups include the Iranians and the Turks. Smaller groups include the Greek and Turkish Cypriots, and the Jews in Israel.

Seven countries in the region are monarchies: Jordan and Saudi Arabia are headed by kings, while emirs rule Bahrain, Kuwait, Qatar and United Arab Emirates. The head of state in Oman is the sultan. The other countries are republics. They include Iran, an Islamic republic, whose laws are based on the teachings of Islam.

Since 1974, Cyprus has been divided into the Greek Cypriot Republic in the south and the so-called Turkish Republic of Northern Cyprus, recognized only by Turkey. The split followed fighting between the island's peoples. Conflict has also occurred in Israel, where the Israelis have battled for survival against their Arab neighbours.

▲ Arab rulers
The kings, emirs and sultans of the region's seven monarchies have considerable power. They govern with prime ministers.

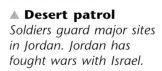

▲ Desert patrol
Soldiers guard major sites in Jordan. Jordan has fought wars with Israel.

▶ Grapefruits, Cyprus
Citrus fruits, such as grapefruits, oranges and lemons, grow well in the sunny countries of Cyprus, Iran and Israel.

▲ Gateway to a souk
Damascus, Syria's capital, has a large souk (market). Visitors may get lost in its narrow streets.

▼ Beach resort, Turkey
Tourism has taken off in Turkey, which has beautiful sunny beaches and many fascinating historic ruins.

▲ Arabian spices
Many spices are used in Arab cooking.

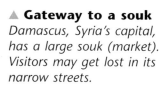

FACTS

ISRAEL
Medinat Israel – State of Israel
Area: *21,056 sq km*
Population: *5,692,000*
Capital: *Jerusalem*
Other cities: *Tel Aviv-Jaffa, Haifa*
Highest point: *Mount Meron (1,208 m)*
Official languages: *Hebrew, Arab*
Currency: *Shekel*

JORDAN
Al-Mamlaka Al-Urduniya Al-Hashemiyah – Hashemite Kingdom of Jordan
Area: *97,740 sq km*
Population: *4,312,000*
Capital: *Amman*
Other cities: *Zarqa, Irbid*
Highest point: *Jabal Ramm (1,754 m)*
Official language: *Arabic*
Currency: *Jordan dinar*

KUWAIT
Dowlat al Kuwait – State of Kuwait
Area: *17,818 sq km*
Population: *1,590,000*
Capital: *Kuwait*
Other cities: *Al Jahra, Salimiya*
Highest point: *283 m*
Official language: *Arabic*
Currency: *Kuwaiti dinar*

LEBANON
Jumhouriya al-Lubnaniya – Republic of Lebanon
Area: *10,400 sq km*
Population: *4,079,000*
Capital: *Beirut*
Other cities: *Tripoli, Sidon*
Highest point: *Qurnat as Sawda (3,083 m)*
Official language: *Arabic*
Currency: *Lebanese pound*

▼ **Turkish coppersmith**
Beautiful metal objects, carpets and highly decorated dishes and bowls are made by Turkish craftworkers and sold in souks (markets).

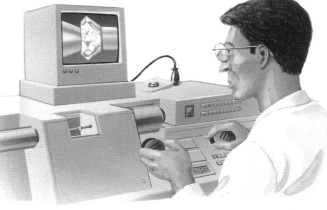

▲ **Diamond polishing**
Israeli craftsmen cut and polish diamonds which are exported to many countries.

▲ **Yemeni market-**
Many goods are on sale at markets in San'a, Yemen.

Several Southwest Asian countries are rich in oil, the region's chief natural resource. Exports of oil have helped several countries to raise the standards of living of the people. Saudi Arabia is the chief oil producer and it has the world's largest reserves. Crude oil is refined to make many useful exports including petrol and petro-chemicals.

Two-thirds of the people of Southwestern Asia live in cities. The largest cities include Istanbul and Ankara in Turkey, Tehran in Iran, Baghdad in Iraq, and Damascus in Syria. Many cities contain old quarters, with narrow roads and large markets called bazaars or souks, alongside modern skyscrapers.

About a third of the people live in villages and work as farmers. Others graze herds of camels, goats and sheep. Major crops, often grown on irrigated land, include barley and wheat, while dates, olives, nuts and citrus fruits are also important. The chief manufacturing countries are Iran and Turkey.

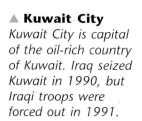

▲ **Kuwait City**
Kuwait City is capital of the oil-rich country of Kuwait. Iraq seized Kuwait in 1990, but Iraqi troops were forced out in 1991.

▲ **Oil well in Iraq**
Southwest Asia has about two-thirds of the world's known oil reserves.

◄ **Oil refinery, Saudi Arabia**
Refineries process oil, producing fuels and chemicals that are used to make many products.

FACTS

OMAN
Sultanat 'Uman – Sultanate of Oman
Area: *212,457 sq km*
Population: *2,173,000*
Capital: *Muscat*
Other cities: *Salalah, Sur*
Highest point: *Jabal Ash Sham (3,035 m)*
Official language: *Arabic*
Currency: *Omani rial*

QATAR
Dawlat Qatar – State of Qatar
Area: *11,000 sq km*
Population: *658,000*
Capital: *Doha*
Other cities: *Dukhan, Umm Said*
Highest point: *103 m*
Official language: *Arabic*
Currency: *Qatar riyal*

SAUDI ARABIA
Mamlaka al-'Arabiya as Sa'udiya – Kingdom of Saudi Arabia
Area: *2,149,690 sq km*
Population: *19,409,000*
Capital: *Riyadh*
Other cities: *Jiddah, Mecca*
Highest point: *Jabal Sawda (3,133 m)*
Official language: *Arabic*
Currency: *Saudi riyal*

SYRIA
Jumhuriya al-Arabya as-Suriya – Syrian Arab Republic
Area: *185,180 sq km*
Population: *14,502,000*
Capital: *Damascus*
Other cities: *Aleppo, Homs*
Highest point: *Mount Hermon (2,814 m)*
Official language: *Arabic*
Currency: *Syrian pound*

Southwest Asia

THE MOST NUMEROUS PEOPLE IN SOUTHWEST ASIA ARE the Arabs. Their language, Arabic, is the only official language in 11 of the 15 countries. Arabic is also an official language in Israel, together with Hebrew which is spoken by the largest group of people, the Jews. The other main languages in Southwest Asia are Farsi (also called Persian), which is spoken in Iran, Greek in Cyprus, and Turkish in Turkey and Cyprus.

Southwest Asia was the birthplace of three great religions: Judaism, which is now the chief religion in Israel; Christianity, which is important in Cyprus and Lebanon; and Islam. About 86 per cent of the people of Southwest Asia follow Islam, the religion of the Arabs.

The region is known for its Islamic art, which includes many arts and crafts, such as book bindings and illustrations, calligraphy (beautiful writing), carvings, ceramics, glassware, metalware, rugs and textiles. Islamic art reaches it finest form in architecture, especially the superbly decorated mosques found throughout the region. The decorations use abstract patterns, often using the shapes of leaves or winding plant stems. Unlike Christian churches, the mosques do not show pictures of people or animals, because they are forbidden by Islamic leaders.

▲ **Wailing Wall**
Jews visit the Western, or Wailing, Wall of an ancient temple in Jerusalem, Israel, to pray.

▲ **Covered faces**
Many Muslim women use veils called yashmaks to hide all but their eyes. In some countries, working women do not wear veils.

▼ **Islamic art**
Abstract patterns and Arabic script are used to decorate mosques and other Islamic buildings. Islam forbids the depiction of people.

◄ **Mosque in Qom**
Qom is a holy city in Iran. Muslim traditions are strictly enforced in Iran. Its mosques are among the most beautiful in the world.

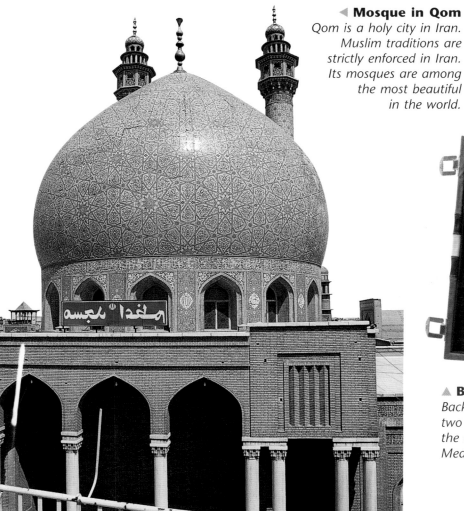

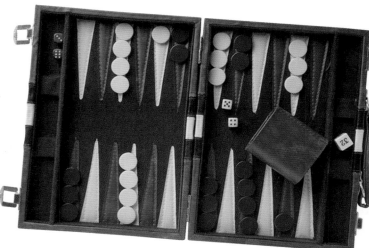

▲ **Backgammon board**
Backgammon is a board game for two players. It is especially popular in the countries bordering the eastern Mediterranean Sea.

▲ Ring-shaped bread rolls
Israelis enjoy bread rolls, called bagels. Other Israeli foods include chicken soup, chopped liver and felafel (ground, fried chickpeas flavoured with onion and spices).

▲ Whirling dervishes
Some Muslim mystics, called dervishes, whirl and dance as part of their religious worship. The order of dancing dervishes was founded in 1273.

▲ Donkey ride
Donkeys and camels are used to transport people in many country areas in Southwestern Asia.

▼ Living together
Kibbutz is the name for a Jewish community in Israel where no one owns property. The people share work, goods and services.

Islam plays an important part in daily life. In Iran, women wear full-length, black body veils, called chadars. They cover their ordinary clothes, concealing their head, shoulders and, usually, the lower part of their faces. But in some Muslim countries, women wear western clothes. Women were once confined to their homes, but today many work in business, education and government.

Basketball, soccer, weight-lifting and wrestling are popular sports in many areas, while many people enjoy such board games as backgammon and chess. Skiing is popular in mountain areas and camel-racing is another sport in the Arabian peninsula.

Flat bread and rice are basic foods in Southwest Asia and dates are important in desert countries. Kebabs of pieces of meat and vegetables cooked on skewers are popular, as are dairy products, such as cheese and yogurt.

▲ Marsh Arabs
The marshy lower valleys of the Tigris and Euphrates rivers in southern Iraq are home to a people called the Marsh Arabs.

Southwest Asia

SOUTHWEST ASIA PLAYED A LEADING ROLE IN THE development of world civilization. It was there, about 11,000 years ago, that people first began to farm the land. Later people built the world's first towns and cities. They also invented writing and devised systems of laws.

Great civilizations developed in the lands around the Tigris and Euphrates rivers, a region called Mesopotamia, which means 'between the rivers'. The major civilizations were those of the Sumerians, Babylonians and Assyrians. The Persians conquered Mesopotamia in 539BC, but the Persian Empire fell to the Greek general Alexander the Great in 331BC.

The Romans ruled much of Southwest Asia at the time when Jesus Christ was born. Muhammad, the founder of Islam, was born in Mecca in the Arabian peninsula in about AD570. His followers later built up a great Arab empire. Later, the Turks ruled much of Southwest Asia, but, after years of decline, the Turkish Ottoman Empire collapsed at the end of World War I. After the war, several new Arab nations were created.

◄ Ancient Sumerian
Sumer, in southeast Iraq, was the world's first major civilization (around 3000–2000BC). Its people produced beautiful cloth, jewellery, pottery and metalwork.

▼ Roman city
The Roman city of Palmyra stands in the desert in central Syria.

▲ Trojan horse, Turkey
According to legend, during the Trojan War around the 12th century BC, Greek soldiers entered the city of Troy by hiding inside a huge, wooden horse.

FACTS

TURKEY
Türkiye Cumhuriyeti – Republic of Turkey
Area: *774,815 sq km*
Population: *62,697,000*
Capital: *Ankara*
Other cities: *Istanbul, Izmir*
Highest point: *Mount Ararat (5,185 m)*
Official language: *Turkish*
Currency: *Turkish lira*

UNITED ARAB EMIRATES
Imarat al-Arabiya al-Muttahida – United Arab Emirates
Area: *83,600 sq km*
Population: *2,532,000*
Capital: *Abu Dhabi*
Other cities: *Dubai, Sharjah*
Highest point: *Jabal Yibir (1,527 m)*
Official language: *Arabic*
Currency: *Dirham*

YEMEN
Jamhuriya al Yamaniya – Republic of Yemen
Area: *527,968 sq km*
Population: *15,778,000*
Capital: *San'a*
Other cities: *Aden, Ta'izz*
Highest point: *Mount Hadur Shuayb (3,760 m)*
Official language: *Arabic*
Currency: *Riyal*

▲ Hittite soldier
The Hittites of ancient Turkey conquered much of southwest Asia between 1900BC and 1200BC.

▼ Crusaders' castles
The Crac des Chevaliers (Castle of the Knights) is one of several medieval castles in Syria. It was an important base for the Christian Crusaders from 1142 to 1271.

▼ Muslim warrior
Saladin (c.1137–93) was sultan of Egypt, Syria, Palestine and Yemen. A brave soldier, he led the Muslims against the Crusaders. He captured Jerusalem in 1187.

◄ Arabian hero
TE Lawrence (1888–1935), a British soldier, helped to organize an Arab revolt against the Turks during World War I. As a result, he became known as Lawrence of Arabia

► Golda Meir
Meir (1898–1978) was Israel's prime minister in 1974 when Israel fought the Yom Kippur War against Arab forces.

▲ Revolution in Iran
In 1979, Ruhollah Khomeini (c.1900–89), a Muslim ayatollah (religious leader), overthrew Iran's king. He made Iran an Islamic republic. He ruled Iran for 10 years.

In the 1930s, many Jewish refugees settled in Palestine in the eastern Mediterranean region. They founded the State of Israel in 1948 and defeated Arab forces in a war which ended in 1949. Further Arab-Israeli wars occurred in 1956, 1967 and 1974. During the wars, Israel occupied the Sinai peninsula and the Gaza Strip from Egypt, the West Bank from Jordan, and the Golan Heights from Syria. In 1979, Israel and Egypt signed a peace treaty and Egypt regained the Sinai peninsula.

◄ Kurdish refugees
The Kurdish people, who live in parts of Armenia, Iran, Iraq, Syria and Turkey, want to set up their own country. Fighting in the area has forced many people to flee from their homes.

In the 1990s, Israel and the PLO (Palestine Liberation Organization) led by Yasir 'Arafat (b.1929) discussed the creation of a new state for Palestinians, including the Gaza Strip and part of the West Bank.

Southwest Asia was also the scene of other conflicts. The Kurds fought unsuccessfully to found a Kurdish state and Iran and Iraq fought a border war in the 1980s. In 1990, Iraq invaded Kuwait. But an international force led by the United States forced the Iraqi president Saddam Hussein (b.1937) to withdraw his troops in 1991.

▲ War damage, Lebanon
Fighting between Christians, Muslims, Palestinian refugees and Syrians occurred in Lebanon between the 1970s and 1991.

◄ Peace talks
In 1993, Israel's prime minister Yitzhak Rabin (1922–95) and Palestinian leader Yasir 'Arafat worked to produce peace in Israel. They were both awarded the Nobel Peace Prize in 1994.

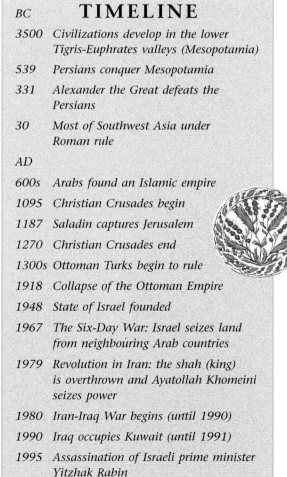

BC	**TIMELINE**
3500	Civilizations develop in the lower Tigris-Euphrates valleys (Mesopotamia)
539	Persians conquer Mesopotamia
331	Alexander the Great defeats the Persians
30	Most of Southwest Asia under Roman rule
AD	
600s	Arabs found an Islamic empire
1095	Christian Crusades begin
1187	Saladin captures Jerusalem
1270	Christian Crusades end
1300s	Ottoman Turks begin to rule
1918	Collapse of the Ottoman Empire
1948	State of Israel founded
1967	The Six-Day War: Israel seizes land from neighbouring Arab countries
1979	Revolution in Iran: the shah (king) is overthrown and Ayatollah Khomeini seizes power
1980	Iran-Iraq War begins (until 1990)
1990	Iraq occupies Kuwait (until 1991)
1995	Assassination of Israeli prime minister Yitzhak Rabin

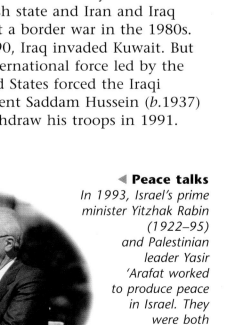

Southern Asia

► Yak

▲ **The Ganges: a holy river**
Hindu pilgrims visit holy Indian cities, such as Varanasi, to bathe in the sacred Ganges River.

Southern Asia lies between Iran to the west and Myanmar to east. China lies to the north, while Tajikistan, Uzbekistan and Turkmenistan border Afghanistan in the far northwest.

The northern part of the region is mountainous. It includes such great ranges as the Hindu Kush in Afghanistan, the Karakoram Range between Pakistan and China, and the mighty Himalayas. Fertile plains extend from eastern Pakistan across northern India into Bangladesh. Some of the world's greatest rivers, including the Indus, Ganges and Brahmaputra, drain these plains. The lower parts of the Brahmaputra and Ganges in Bangladesh have created the world's largest delta.

▲ **The Himalayas**
The Himalayas is the highest mountain range in the world. It includes Mount Everest which rises to 8,848 metres above sea level.

Southern India consists mainly of a large plateau called the Deccan, which is bordered by two mountain ranges, the Western and Eastern Ghats. Sri Lanka contains a mountain region surrounded by fertile plains. The Maldives consists of a chain of low coral islands lying south of India.

The northwest is mostly dry and includes the Thar Desert on the border between Pakistan and India. The northern mountains have cold winters, but much of Southern Asia is warm and tropical.

The heavy rains that fall in summer are the main feature of the climate. The rains are brought by moist, monsoon winds that blow from the sea between May and October. Farmers eagerly await the rains, but sometimes they make the rivers overflow, causing widespread flooding. For example, the rivers in Bangladesh burst their banks in 1998. More than two-thirds of the country was under water.

◄ **Wind-surfing**
Many tourists visit the Maldives, an island group south of India. The islands have a sunny climate and beautiful beaches.

FACTS

AFGHANISTAN
Islamic Emirate of Afghanistan
Area: *652,090 sq km*
Population: *24,167,000*
Capital: *Kabul*
Other cities: *Qandahar, Herat, Mazar-e-Sharif*
Highest point: *Nowshak (7,485 m)*
Official languages: *Pashto, Dari (Persian)*
Currency: *Afghani*

BANGLADESH
Gana Prajatantri Bangladesh – People's Republic of Bangladesh
Area: *143,998 sq km*
Population: *121,671,000*
Capital: *Dhaka*
Other cities: *Chittagong, Khulna*
Highest point: *Mount Keokradong (1,230 m)*
Official language: *Bengali*
Currency: *Taka*

BHUTAN
Druk-yul – Kingdom of Bhutan
Area: *47,000 sq km*
Population: *715,000*
Capital: *Thimphu*
Other cities: *Phuntsholing*
Highest point: *Kula Kangri (7,554 m)*
Official language: *Dzonghka*
Currency: *Ngultrum*

INDIA
Bharat – Republic of India
Area: *3,287,590 sq km*
Population: *945,121,000*
Capital: *New Delhi*
Other cities: *Mumbai, Calcutta, Chennai*
Highest point: *Kanchenjunga (8,598 m)*
Official language: *Hindi, English*
Currency: *Indian rupee*

◀ Mountain country
Leh is a town in northwestern India, in the western part of the Himalaya range.

▲ Hindu architecture
The Teli-ka-Mandir is an ancient Hindu temple. It is in the city of Gwalior, central India.

AFGHANISTAN

PAKISTAN

▲ Nepalese temple
Nyatapola temple is the highest temple in Nepal. It is in the town of Bhaktapur, a short distance to the east of the capital, Katmandu.

MALDIVES

NEPAL

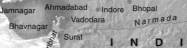  

BHUTAN

BANGLADESH

▼ Indian rhino
Like the Bengal tiger, the Indian rhinoceros is now rare.

INDIA

SRI LANKA

▼ Reclining Buddha
The ancient Sri Lankan capital of Polonnaruwa has huge carvings of the Buddha. Buddhism is the chief religion in Sri Lanka.

▶ Tricky terrain
Travel is difficult in the mountains of northern Pakistan. Rope bridges span many of the rivers.

Map labels:

TURKMENISTAN
TAJIKISTAN
Mazar-e-Sharif
Herat
HINDU
DISPUTED AREA
K2 8,611m
KARAKORAM
AFGHANISTAN
Kabul
Farah
Khyber Pass
Peshawar
Srinagar
Islamabad
Qandahar
Rawalpindi
JAMMU & KASHMIR
RIGESTAN DESERT
Quetta
Faisalabad
Lahore
Amritsar
SULAIMAN RANGE
Multan
Sutlej
PUNJAB
PAKISTAN
Indus
Bahawalpur
Nanda Devi 7,817m
BALUCHISTAN PLATEAU
Sukkur
Delhi
Tibet (CHINA)
Hyderabad
GREAT INDIAN DESERT (THAR DESERT)
New Delhi
NEPAL
Mt Everest 8,848m
Thimphu
Karachi
Jodhpur
Jaipur
Bareilly
Annapurna 8,078m
Gulf of Kachch
Ajmer
Agra
Lucknow
Katmandu
BHUTAN
HIMALAYA
Udaipur
Kota
Gwalior
Kanpur
Ghagara
Brahmaputra
Gauhati
NAGA HILLS
Allahabad
Varanasi
Patna
Jamnagar
Ahmadabad
Indore
Bhopal
Son
Ganges
BANGLADESH
Imphal
Bhavnagar
Vadodara
Narmada
Jabalpur
Asanol
Dhaka
MYANMAR (BURMA)
Surat
Gulf of Khambhat
Jamshedpur
Khulna
Chittagong
INDIA
Calcutta
Mumbai (Bombay)
Aurangabad
Nagpur
Raipur
Mahanadi
Mouths of the Ganges
Pune
DECCAN
Cuttack
WESTERN GHATS
Godavari
Solapur
Kolhapur
Hyderabad
Krishna
Vishakhapatnam
EASTERN GHATS
Hubli-Dharwar
Kurnool
Vijayawada
Penner
Nellore
Mangalore
Bangalore
Chennai (Madras)
Mysore
Kozhikode
Coimbatore
Tiruchchirappalli
Cochin
Madurai
Palk Strait
Jaffna
Trivandrum
Trincomalee
C. Comorin
Gulf of Mannar
SRI LANKA
Colombo
Kandy
Pidurutalagala 2,524m
Galle
MALDIVES
INDIAN OCEAN
N E S W

167

Southern Asia

SOUTHERN ASIA CONTAINS EIGHT COUNTRIES.
The largest is India, which makes up nearly two-thirds
of the region and contains about three-quarters of
the people. Pakistan, the second-largest country,
makes up more than 15 per cent of the region.

India is a federal state, made up of 25 states,
each with its own government, and seven
territories. Bangladesh, the Maldives, Pakistan
and Sri Lanka are also republics. Afghanistan was
a republic until 1997, when its government, which
was engaged in a civil war, declared the country to
be an Islamic Emirate. Southern Asia also contains
two kingdoms: Bhutan and Nepal.

Since 1947, when modern India and Pakistan
were born, some of their boundaries have been
disputed. The chief problem concerns a region in
northwestern India, called Jammu and Kashmir.
India claims the entire area, but Pakistan occupies
part of it.

▲ **Basket-making**
Indian basket-makers use
natural fibres, such as the leaves
of palm trees.

▲ **Elephant rides**
*Painted elephants carry
visitors up the steep
slopes to the Indian fort
of Amber, near Jaipur.*

▲ **Bullock cart**
*Farmers still use
traditional forms
of transport in
Sri Lanka and
other countries
in Southern Asia.*

▲ **Sandal maker**
*Footwear and clothes are still made
by hand in Southern Asia. But
there are also huge factories which
manufacture items for export.*

◀ **Tea leaves**
*The leaves of tea plants are so
delicate they must be hand-picked.
India and Sri Lanka are among the
world's leading tea producers.*

▼ **Growing rice**
*Paddy (or rice) fields, enclosed
by earth banks, are ploughed
and flooded before rice shoots
are planted.*

FACTS

MALDIVES
*Divehi Raajjeyge Jumhooriyya –
Republic of the Maldives*
Area: *298 sq km*
Population: *256,000*
Capital: *Malé*
Other cities: *none*
Highest point: *Wilingili Island
(24 m)*
Official language: *Divehi*
Currency: *Rufiyaa*

NEPAL
Nepal Adhirajya – Kingdom of Nepal
Area: *147,181 sq km*
Population: *22,037,000*
Capital: *Katmandu*
Other cities: *Lalitpur, Biratnagar*
Highest point: *Mount Everest
(8,848 m)*
Official language: *Nepali*
Currency: *Nepalese rupee*

PAKISTAN
*Islami Jamhuriya e Pakistan –
Islamic Republic of Pakistan*
Area: *796,095 sq km*
Population: *133,510,000*
Capital: *Islamabad*
Other cities: *Karachi, Lahore,
Faisalabad*
Highest point: *K2 (8,611 m)*
Official language: *Urdu*
Currency: *Pakistan rupee*

SRI LANKA
*Sri Lanka Prajathanthrika Samajavadi
Janarajaya – Democratic Socialist
Republic of Sri Lanka*
Area: *65,610 sq km*
Population: *18,502,000*
Capital: *Colombo*
Other cities: *Dehiwela-Mt Lavinia,
Moratuwa*
Highest point: *Pidurutalagala
(2,524m)*
Official languages: *Sinhala, Tamil*
Currency: *Sri Lanka rupee*

◄ Pink palace
The Hawa Mahal, or the Palace of the Winds, is in the beautiful city of Jaipur. Jaipur was founded in 1728.

◄ Rickshaw
A rickshaw is a two-wheeled carriage pulled by a man. The pedicab, drawn by a man on a bicycle, is also known as a rickshaw. These forms of transport compete for passengers with taxis in many cities.

▲ Computer industry
Factories in Bangalore produce computers. High-tech industry is increasing in India, though most people still live by farming.

▼ Fishing in Bangladesh
Fish are caught in the Bay of Bengal and also in the many waterways that drain the flat land.

Most people in Southern Asia are poor. Shattered by civil war, Afghanistan is one of the world's poorest countries; and more than 90 per cent of the people in Bhutan and Nepal are poor farmers. Farming is the chief activity in Southern Asia, employing about 63 per cent of the people. The chief crop is rice. India ranks second only to China in world production, while Bangladesh ranks fourth.

Other food crops include barley, millet, sorghum and wheat, together with such vegetables as beans and peas, coconuts, a variety of fruits and tea. Cotton, jute, rubber and tobacco are also grown. Livestock farming is important in some countries, but Hindus do not eat meat. However, Hindus drink milk from buffaloes and cattle.

▲ Chilli peppers

India has mineral reserves, including coal, iron ore and oil. Manufacturing has increased rapidly in the last 50 years, especially in the cities. The largest cities in India are Mumbai (formerly Bombay), Delhi, Calcutta and Chennai (formerly Madras). Other large cities include Karachi and Lahore in Pakistan. Many cities of Southern Asia contain old areas with slums, alongside modern high-rise buildings. The contrasts between the lives of rich and poor people are very great.

► Indian market
Many Indians do not eat meat. They buy vegetables and spices, which they eat with rice or bread.

Southern Asia

INDIA'S PEOPLE ARE DIVIDED INTO TWO MAIN GROUPS: the lighter-skinned Indo-Aryans in the north and the darker-skinned Dravidians in the south. India has 16 major languages and more than 1,000 minor languages and dialects. The main official language is Hindi. English is an 'associate' national language, spoken by most educated Indians.

The people in the northwestern part of Southern Asia include Afghans, Indo-Aryans, Persians and others descendants of Southwestern Asian people. The Sinhalese of Sri Lanka and the people in the Maldives are descendants of people from northern India, while the ancestors of Sri Lankan Tamils came from southern India.

Southern Asia's arts include great epic poems and magnificent architecture and sculpture, such as superb Hindu temples, Muslim mosques, and great statues of the Buddha. Other arts include traditional music and dance.

Many people in the cities of Southern Asia wear Western clothes, but many people prefer light, cool clothes. Many men wear dhotis – white cloths wrapped around the body and tucked between the legs, while Indian women often wear colourful saris – pieces of cloth draped around the body like a dress. Many men wear turbans or other kinds of hats to protect them against the sun.

▲ Hindu wedding
Wealthy Indians have elaborate wedding ceremonies. These days, many weddings take place at hotels.

▲ Houseboats
Some people in northwestern India live on boats on the lakes. Today, tourists hire houseboats. They can study at first hand the life of the lake people.

▼ Road transport
Long-distance drivers in India and Pakistan often decorate their trucks. Road transport is important because distances in Southern Asia are so great.

▲ Sherpas in Nepal
The Sherpa people of northeastern Nepal are famous as mountaineers and soldiers. Some work as mountain guides.

▲ Snake charmer
Snake charmers in southern Asia play music and sway to and fro. This makes their cobras rise up and follow their movements.

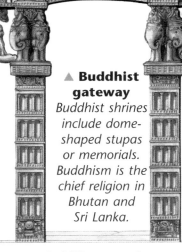

▲ Buddhist gateway
Buddhist shrines include dome-shaped stupas or memorials. Buddhism is the chief religion in Bhutan and Sri Lanka.

▶ Village gods
Eight out of every ten Indians are Hindus. Many villages have statues of gods.

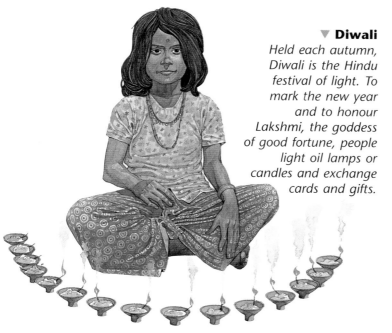

▼ Diwali
Held each autumn, Diwali is the Hindu festival of light. To mark the new year and to honour Lakshmi, the goddess of good fortune, people light oil lamps or candles and exchange cards and gifts.

▲ Jain building
Jainism is an Indian religion dating back 2,500 years. This Jain temple is in the Great Indian (Thar) Desert region in northwestern India.

The people of Southern Asia celebrate many religious festivals, such as Diwali, the Indian festival of light, and Holi, when people throw coloured water and powders on each other. The main food in India is rice, which is eaten with spicy vegetables. People also eat a flat bread called chapati.

▶ Sacred monkeys
Langurs are monkeys found in India. Hanuman langurs are regarded as sacred.

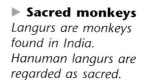

Hinduism, one of the world's oldest religions, is followed by 63 per cent of the people of Southern Asia. Hindus worship many gods, who are forms of one universal spirit. Hindus believe that the soul never dies. Instead, it is reincarnated (reborn) in the body of a human or animal.

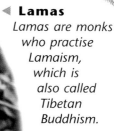

◀ Lamas
Lamas are monks who practise Lamaism, which is also called Tibetan Buddhism.

Hindus make up 80 per cent of India's population, but India also has about 100 million Muslims, 36 million Christians and nearly seven million Buddhists. Hinduism is the main religion in Nepal, while Islam is the religion of Afghanistan, Bangladesh, the Maldives and Pakistan. Buddhism is the leading faith in Bhutan and Sri Lanka.

▲ Prayer bells
Buddhists often shake bells during worship. The sounds are supposed to frighten away evil spirits and attract good ones.

▼ Spring festival
Holi is an ancient Hindu spring festival, when people throw coloured water and powders over each other. Differences of age, caste, sex and status are disregarded on this day.

▼ Temple at Khajuraho
Khajuraho, south of Kanpur in north-central India, is a site containing 22 great temples. The temples contain elaborate carvings.

Southern Asia

BETWEEN 2500BC AND 1700BC, THE INDUS Valley civilization flourished in what is now Pakistan and northwestern India. Around 1500BC, Aryan people from the northwest invaded India and developed the Hindu cultures of the region. After Alexander the Great reached the Indus Valley region around 320BC, a major Buddhist culture grew up in the north. Following other invasions, Mongol invaders took northern India in the AD1400s. By 1526, India was part of a large Mongol empire.

From the 16th century, European traders began to work in Southern Asia. As the Mongol empire declined, the British East India Company gradually gained control of large areas. An unsuccessful rebellion against the East India Company took place in 1857 and, in 1858, Britain took over control of the region. Britain helped to develop the region by building railways, a telephone system and irrigation works.

But many Indians wanted independence. From 1920, they were led by a lawyer, Mohandas K Gandhi (1869–1948), also known as Mahatma Gandhi, who led campaigns of non-violent disobedience. After World War II, Britain agreed to make India independent. Following violence between Hindus and Muslims, the country's leaders agreed to partition British India into Muslim Pakistan and a mainly Hindu India. Millions of people became refugees and many were killed. Fighting also occurred in the northwest, where India and Pakistan fought over Kashmir.

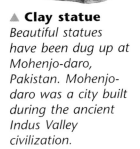

▲ **Clay statue**
Beautiful statues have been dug up at Mohenjo-daro, Pakistan. Mohenjo-daro was a city built during the ancient Indus Valley civilization.

▲ **On the banks of the Indus**
The ancient citadel of Mohenjo-daro had wide streets, large houses and good sewers.

◀ **Emperor Asoka**
Emperor Asoka ruled over the Maurya Empire (272–232BC). He conquered large areas and converted from Hinduism to Buddhism.

▼ **Taj Mahal**
The Taj Mahal, in Agra, India, was built as a tomb for Mumtaz Mahal, favourite wife of Shah Jahan (1592–1666). She died in 1631, but the marble tomb was not completed until 1643.

▲ **Golden Temple, Amritsar**
The Golden Temple, the most sacred Sikh temple, is in Amritsar, in northwestern India. Sikhs make up two per cent of India's population.

▼ National leaders

In 1947, Muslim Pakistan, under Muhammad Ali Jinnah (1876–1948), broke away from India, under Jawaharlal Nehru (1889–1964).

▲ Mass migration

When India and Pakistan became independent in 1947, millions of Muslims and Hindus moved home. Attacks on refugee columns led to many deaths.

Pakistan was divided into two separate areas: West and East Pakistan. In 1971, civil war broke out and East Pakistan became a separate country called Bangladesh. Civil war also occurred in Afghanistan and Sri Lanka. In 1979, Russian troops entered Afghanistan to protect its Communist government, but they withdrew in 1988 and Islamic forces seized control. This civil war continued into the late 1990s.

In Sri Lanka, some Hindu Tamils supported a guerrilla group called the Tamil Tigers to fight for their own country which they want to call Tamil Eelam. ('Eelam' means 'state'.) Meanwhile, relations between India and Pakistan were strained. In 1998, both countries tested nuclear devices. This revealed the danger that nuclear weapons might be used in any future war.

◀ Democratic elections

India is the world's biggest democracy, with more voters than any other country. It has held regular elections since it won independence.

◀ Independence day parade

Soldiers march in impressive parades to celebrate India's independence from Britain. Military service in India is voluntary.

BC	TIMELINE
2500	Indus Valley civilization develops
320s	Alexander the Great reaches southern India
272	Buddhist empire unites much of southern India (until 232)
AD	
1200s	Mongols capture northern India
1526	Mongol empire established in Southern Asia
1619	Portuguese control most of Ceylon (Sri Lanka)
1649	Taj Mahal completed
1757	British East India Company wins control of Bengal
1802	Ceylon becomes a British Crown Colony
1858	Britain governs India
1947	British India gains independence and splits into two countries: India and Pakistan; fighting breaks out in Kashmir
1948	Mahatma Gandhi assassinated by Hindu fanatic
1971	East Pakistan becomes a new country, Bangladesh
1972	Ceylon becomes the Republic of Sri Lanka
1988	USSR withdraws from Afghanistan
1992	Rebels seize Kabul, Afghanistan

Eastern Asia

▶ Panda
Giant pandas live in the bamboo forests of western and southwestern China.

By area, Eastern Asia is dominated by China, the world's third-largest country after Russia and Canada. China claims the island of Taiwan, although Taiwan is a separate territory with its own government, and also the tiny territory of Macao, in the southeast, which Portugal has agreed to return to China in December 1999. Eastern Asia also includes North Korea, South Korea and Mongolia. Japan is also part of Eastern Asia, but is covered separately.

The western part of Eastern Asia contains high plateaux, mountain ranges and deserts. Tibet, in the southwest, contains the region's highest peak, Mount Everest, which stands on the border with Nepal. The inland deserts are hot in summer, but are bitterly cold in winter. The Gobi Desert, which lies partly in northern China and partly in Mongolia, is one of the bleakest deserts on Earth.

▲ Monk, Lhasa
Tibet, whose capital is Lhasa, is a region of China. But many Tibetans would like Tibet to become an independent country in its own right.

By contrast, the eastern part of this region contains fertile plains, uplands and densely-populated river valleys. Three great rivers cross eastern China. They are the Huang He (formerly called the Hwang Ho, or Yellow River), the Chang Jiang (Yangtze) and the Xi Jiang in the southeast. North and South Korea occupy a rugged peninsula which is attached to northeastern China. Most Koreans live on the plains and around the coast.

Tibet is a high, windswept plateau, with bitterly cold winters and short, cool summers. The north and northwest is mainly dry, with bare desert or areas of dry grassland, called steppe. Summers in southeast China range from warm to hot, while its winters are cool, but the northeast, including most of Korea, has cold winters.

◀ Forest of stone
Worn limestone rocks, resembling giant tree-trunks, form the Yunnan stone forest in southern China. The area lies southeast of Kunming.

FACTS

CHINA
Zhonghua Renmin Gonghe Guo – People's Republic of China
Area: *9,596,961 sq km*
Population: *1,193,951,000*
Capital: *Beijing*
Other cities: *Shanghai, Tianjin, Shenyang, Wuhan*
Highest point: *Mount Everest (8,848 m)*
Official language: *Mandarin Chinese*
Currency: *Yuan*

MACAO
Portuguese territory, also called Macau
Area: *18 sq km*
Population: *461,000*
Capital: *Macao*
Other cities: *none*
Highest point: *Coloane Alto (174 m)*
Official language: *Chinese, Portuguese*
Currency: *Pataca*

MONGOLIA
Mongol Uls – Mongolia
Area: *1,566,500 sq km*
Population: *2,516,000*
Capital: *Ulan Bator*
Other cities: *Darhan, Erdenet*
Highest point: *Altai Mountains (4,362 m)*
Official language: *Khalka Mongolian*
Currency: *Tugrik*

NORTH KOREA
Chosun Minchu-chui Inmin Konghwa-guk – People's Democratic Republic of Korea
Area: *120,528 sq km*
Population: *22,451,000*
Capital: *Pyongang*
Other cities: *Hamhung, Cho'ngjin*
Highest point: *Paektu-san (Paektu Mountain, 2,744 m)*
Official language: *Korean*
Currency: *Won*

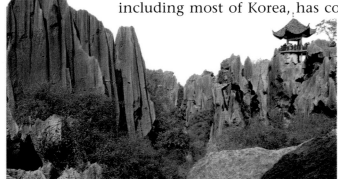

▶ **Seoul, Korea**
Seoul is the capital of South Korea. It is one of the world's largest cities and it has many high-rise buildings and industries.

▲ **Great Wall of China**
The 7,400-km-long Great Wall of China was built to hold back

MONGOLIA

NORTH KOREA

TAIWAN

SOUTH KOREA

HONG KONG

CHINA

Map labels:
MONGOLIA · RUSSIA · KAZAKHSTAN · KYRGYZSTAN · INDIA · NEPAL · BHUTAN · MYANMAR (Burma) · LAOS · VIETNAM · CHINA · NORTH KOREA · SOUTH KOREA · TAIWAN · Hong Kong · MACAO

Hovsgol Lake · Ulaangom · Darhan · Edernet · Ulan Bator · Choybalsan · Tamsagbulag · Qiqihar · Harbin · Mudanjiang · Changchun · Jilin · Ch'ongjin · Tonghua · Fushun · Shenyang · Anshan · Hamhung · Chifeng · Jinzhou · P'yongyang · Wonsan · Kaesong · Seoul · Hovd · Fuhai · Karamay · Ebinur Hu · Yining · Kuytun · Ürümqi · Hami · Dalardzadgad · Baotou · Beijing · Tangshan · Bo Gulf · Dalian · Korea Bay · Weihai · Yantai · Pusan · Kwangju · Aksu · Kashi · Bosten Lake · Turfan Depression · Shizuishan · Yinchuan · Tianjin · Shijiazhuang · Zibo · Qingdao · Cheju I. · Mt. K2 · Hotan · TAKLIMAKAN DESERT · Yumen · Xining · Lanzhou · Taiyuan · Jinan · Huang He · Xuzhou · Hongze Lake · Nantong · Qinghai Lake · Xi'an · Zhengzhou · Nanjing · Shanghai · Tangra Lake · Siling Lake · Nam Lake · Lhasa · Xigaze · Mt. Everest 8,848 m · Qamdo · Chengdu · SICHUAN BASIN · Chongqing · Chao Lake · Hangzhou · Ningbo · Linhai · Macheng · Yichang · Wuhan · Dongting Lake · Poyang Lake · Nanchang · Wenzhou · Leshan · Luzhou · Changsha · Fuzhou · Taipei · Hengyang · Zhangzhou · Xiamen · Kaohsiung · Guiyang · Liuzhou · Xi Jiang · Guangzhou · Shantou · Xiaguan · Kunming · Nanning · Hong Kong · Gejiu · Pingxiang · Zhanjiang · Haikou · Hainan · Gulf of Tongkin

Mountains and regions: ALTAI MTS. · HANGAYN MTS. · HENTYN MTS. · GREATER HINGGAN · LESSER HINGGAN · GOBI DESERT · MU US DESERT · TIAN SHAN · Dzungaria · KUNLUN SHAN · ALTUN SHAN · QILIAN SHAN · KARAKORAM · HIMALAYA · PLATEAU OF TIBET · TANGGULA SHAN · BAYAN HAR SHAN · DALOU SHAN · NAN LING MTS. · AILAO MTS. · Amur · Chang Jiang · Huang He · Salween · Mekong · Taiwan Strait · Korea Bay · YELLOW SEA · EAST CHINA SEA · Bo Gulf

Compass: N E S W

▼ **Shanghai**
Shanghai is China's largest city. It is also the country's chief port and industrial city.

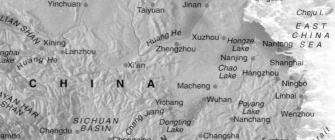

▶ **Hong Kong**
Formerly a British colony, Hong Kong became a Special Administration Region of China in 1997.

▶ **Pekingese dog**

Eastern Asia

IN RECENT YEARS, EASTERN ASIA HAS SUFFERED from many disputes. One issue concerns Taiwan and other offshore islands. Since 1949, when the Chinese Communists took over the mainland, Taiwan has been the home of the Nationalists, who founded a rival 'Republic of China'. China still claims Taiwan, but the Taiwanese oppose union with Communist China. Some problems have been solved. First, in 1997, Britain returned Hong Kong to China. Also, Portugal agreed in 1987 to return Macao, another small territory in southwestern China, in December 1999. Further, China and Russia ended their border disputes in 1998.

But no solution appears in sight concerning Tibet in the southwest. Tibet has been part of China since the early 1950s, but many Tibetans want their independence and the return of their religious ruler, the Dalai Lama, who lives in exile.

▲ Fishing with birds
Chinese fishermen train birds called cormorants to catch fish for them in inland waters.

▲ Golden lion
Beijing's Forbidden City is a large area containing palaces of former emperors. It contains many statues.

FACTS

SOUTH KOREA
Daehan Min-Kuk – Republic of Korea
Area: *99,274 sq km*
Population: *45,545,000*
Capital: *Seoul*
Other cities: *Pusan, Taegu*
Highest point: *Halla San (Halla Mountain, 1,950 m)*
Official language: *Korean*
Currency: *Won*

TAIWAN
Chung-hua Min-kuo – Republic of China
Area: *36,179 sq km*
Population: *21,463,000*
Capital: *Taipei*
Other cities: *Kao-hsiung, T'ai-chung*
Highest point: *Yü Shan (Mount Morrison, 3,997 m)*
Official language: *Mandarin Chinese*
Currency: *New Taiwan dollar*

▶ Kowloon, Hong Kong
Hong Kong consists of 235 islands and an area on the mainland of China. Kowloon is on the mainland. It is a busy place, with many shops and industries.

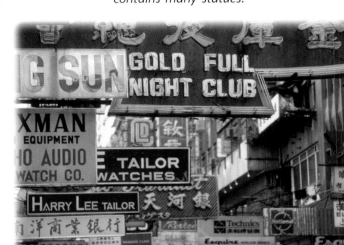

▶ Chinese junk
Junks are wooden vessels with two or more sails. They are used by Chinese and other sailors in Eastern Asia. Some people live on junks.

▼ **Stock exchange, Hong Kong**
*Hong Kong is one of the world's
leading financial centres.*

▲ **Market, Beijing**
*Chinese people from
the countryside sell their
produce at markets in
the cities.*

▼ **Rice
plant**

The border between North and South Korea is disputed. After World War II,
North Korea became a Communist country, while South Korea, aided by the
United States, allied itself with Western countries. In 1950–53, the Korean War
caused great destruction and North and South Korea remain divided. South
Korea is a multi-party democracy, as is Mongolia, which was a Communist
country until 1990. Communist governments still rule North Korea and China.

▲ **Jade camel**
*Since ancient times, the
Chinese have mined jade,
a white or green stone, and
used it to make beautiful
carvings. This jade camel is
more than 1,000 years old.*

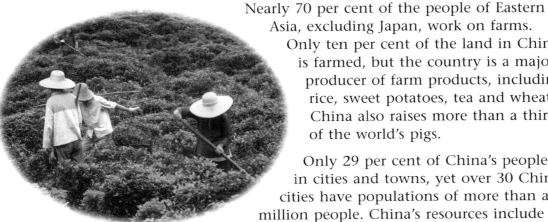

Nearly 70 per cent of the people of Eastern
Asia, excluding Japan, work on farms.
Only ten per cent of the land in China
is farmed, but the country is a major
producer of farm products, including
rice, sweet potatoes, tea and wheat.
China also raises more than a third
of the world's pigs.

Only 29 per cent of China's people live
in cities and towns, yet over 30 Chinese
cities have populations of more than a
million people. China's resources include coal
and oil, and, although many people are poor, it is
a rapidly developing country. China's 'special economic
zones' in the east have many industries, which are partly
financed by foreign companies.

▲ **Picking tea**
*Tea is one of the
main crops in
southern China.
China ranks second
only to India in the
production of tea.*

▼ **Boat builders**
*China has one of the world's
largest fishing industries. About
three-fifths of the catch comes
from the sea.*

South Korea and Taiwan are also
important manufacturing nations.
Communist North Korea also has
industries, though living standards
are lower than those in South Korea.

◄ **Printing in South Korea**
*The printing of books, newspapers and
magazines is important in the drive to
educate people. Most Asians believe that
education is vital if their countries are to
make economic progress.*

Eastern Asia

CHINA CONTAINS MORE THAN ONE-FIFTH OF THE world's population. Until recently, its birth-rate was so high that the government made laws to slow down the rate of population growth. For example, it encourages couples to have one child only. Women may not marry until the age of 20 and men may not marry until they are 22.

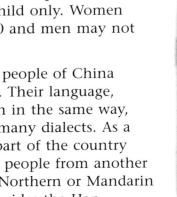

About 92 per cent of the people of China belong to the Han group. Their language, Chinese, is always written in the same way, but spoken Chinese has many dialects. As a result, people from one part of the country often cannot understand people from another part. One dialect, called Northern or Mandarin Chinese, is the official language. Besides the Han, China also has 55 minority groups, some of whom speak Chinese dialects, while others have their own languages. Important minorities include Kazaks and Uigurs in the northeast, Mongols in the north, Koreans in the northeast, and Tibetans in the southeast.

▲ Chinese writing
Chinese writing consists of characters that stand for words or parts of words.

▲ Tiananmen Square, Beijing
Firework displays and parades are held in Tiananmen Square in central Beijing. The Forbidden City and parliament buildings border the square.

▼ Mongolian yurt
The traditional homes of Mongolian herders are felt tents, called yurts. They can be easily taken down and moved to a new site.

▲ Summer festival
Every May, dragon boat races are held in Hong Kong. After the races, there are displays of martial arts and street theatre to enjoy. The revellers snack on dragon boat dumplings and roasted pinenuts and firecrackers are let off well into the night.

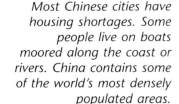

▼ Life on the water
Most Chinese cities have housing shortages. Some people live on boats moored along the coast or rivers. China contains some of the world's most densely populated areas.

▼ Chinese board game
Mah jong developed in the 19th century. It is probably based on games dating back more than 2,000 years.

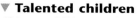

▼ Talented children
Chinese children with special talents receive extra schooling at the weekend. Subjects include learning to play the piba, a stringed instrument.

◄ Buddhist monk
Forms of Buddhism are followed in China, South Korea and Mongolia.

▲ Chinese opera
Beijing opera combines drama with songs and dances. The actors wear elaborate costumes.

The Communist governments in Eastern Asia have discouraged religious worship. However, many people follow Buddhism, Confucianism or Taoism, and sometimes a mixture of all three. Some Muslims live in western China and in Mongolia, while Christians make up around one-fourth of the population of South Korea. Eastern Asia has long artistic traditions. Early Chinese art includes beautiful pottery and carved jade statues. China also has much great literature, painting, sculpture, music and theatre.

The people enjoy many kinds of recreation. Many Chinese get up early every morning to engage in an ancient and graceful form of exercise called taijiquan. Sports include baseball, basketball, soccer, table tennis and volleyball.

Rice is the chief food in the warm, southern parts of the region, while wheat is important in the cooler north. Pork and poultry are leading meats, while bamboo shoots, cabbages and tofu (soya bean curd) are popular vegetables. Beijing duck is a special treat in China. It consists of slices of crisp roast duck wrapped in thin pancakes. Koreans enjoy kimchi, a spicy mixture of cabbage, radishes and other vegetables. Tea is the chief drink in Eastern Asia.

▲ Qing vase
Chinese artists have made pottery for thousands of years. The early part of the Qing dynasty, which began in 1614, produced superb pottery.

◄ Beijing duck
Beijing (or Peking) duck is eaten on special occasions. It consists of slices of roast duck which is eaten with plum sauce, sliced spring onions and thin pancakes.

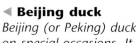

▲ Unarmed combat
Chinese karate is called kung fu. It employs circular movements which differ from the powerful movements of other forms of karate.

◄ Chinese food
Rice is eaten in southern China, but most northerners prefer wheat. Vegetables, pork and poultry are also popular.

179

Eastern Asia

EARLY CIVILIZATIONS DEVELOPED IN THE fertile valleys of eastern China around 3,500 years ago. China's history from the 1700s BC until 1912 is divided into dynasties – that is, rule by a members of a family or group.

The Qin dynasty (221BC–206BC) was the first to set up a strong central government to rule eastern China. The Han dynasty, which followed, founded a strong empire, ruling an area as large as the Roman empire. China came under foreign rule when the Mongols conquered the country in the 13th century and set up the Yuan dynasty. The Mongols conquered a vast area extending westwards from Korea to eastern Europe.

▲ **Terracotta army, Xi'an**
Burial pits near the tomb of Emperor Shi Huangde, who lived in the 3rd century BC, contain lifesize clay statues of soldiers.

▲ **Stone statues**
The road leading to the Ming tombs, just outside Beijing, is lined with impressive statues.

Chinese rule was restored in 1368, but in the 17th century, the Manchu from Manchuria set up the Qin dynasty. Manchu rule was weakened in 1895 when China was defeated by Japan, which occupied Korea in 1910. Dynastic rule came to an end in 1912, when China became a republic. Japan seized Manchuria in 1931 and China fought against Japanese invaders between 1937 and 1945.

At the end of World War II, civil war in China continued between the Communists and the Nationalists. In 1949, the Communists, under Mao Zedong (1893–1976), took over mainland China, while the Nationalists under Chiang Kai-shek (1887–1975) retreated to Taiwan.

▲ **Ming tomb**
Many Chinese and foreigners visit the Ming tombs. The Ming dynasty lasted from 1368 until 1644.

▲ **Silk road traders**
The Silk Road was the trade route in Roman and later times between China and Syria.

▼ **Beijing**
In the Ming dynasty, a wall and moat were built around Beijing. In the Qing dynasty, more palaces, temples and other buildings were added outside the city wall.

▶ **Manchu emperor**
The Manchus set up the Qing dynasty. Only the emperor was allowed to wear a silk gown embroidered with a five-clawed dragon.

Guangzhou (formerly called Canton) is a Chinese port, northwest of Hong Kong. During the 1800s, it was the only Chinese port open to Westerners.

◄ **Dowager Empress**
Cixi ruled as regent of China for two different emperors, her son and her nephew.

▶ **Last emperor**
Pu Yi was China's last emperor. He was two when he became emperor in 1908.

Under Communist rule, the governments of China, North Korea and Mongolia took over all land and industry. Although they improved the lives of many poor peasants, they also suppressed opponents of Communism. Some Communist policies also caused severe economic problems. From the 1980s, China introduced economic reforms, especially in the east, where foreign companies were welcomed. But the Communists kept control of China's government.

▲ **The Long March**
In the 1930s, under Nationalist attack, the Chinese Communists undertook the Long March to northern China. On the march, Mao Zedong emerged as the Communist leader.

On the other hand, Mongolia held free elections in 1990 and this led to the establishment of democratic government. Meanwhile, South Korea, Taiwan and the British territory of Hong Kong set up high-tech and other industries, rapidly developing their economies. In 1997, Hong Kong, which Britain had controlled since 1842, was returned to China, becoming a 'Special Administration Region'. However, the territory retained many of the laws introduced by Britain.

BC	**TIMELINE**
1700s	Eastern Asia's first major civilization founded in the Huang He valley
214	Under Emperor Shih Huang-ti, China is united; Great Wall of China is fortified and joined together
202	AD 220 Chinese culture flourished under the Han dynasty (until 220AD)
AD	
1279	Mongols begin to rule Mongolia, Korea and all of China
1368	Chinese culture developed under the Ming dynasty (until 1644)
1644	Manchus rule China, Mongolia and Korea (until 1910 when Japan occupied the peninsula)
1912	Republic of China founded
1937	China fought a war against Japan (until 1945)
1949	Communists establish the People's Republic of China
1950	Korean War (until 1953)
1989	Tiananmen Square massacre of student demonstrators, Beijing
1997	Britain hands Hong Kong to China
1999	Portugal hands Macao to China

▲ **Chairman Mao**
Mao Zedong (or Mao Tse-tung) became China's leader when the Communists took power in 1949. He died in 1976.

▶ **Deng Xiaoping**
From 1976, Deng Xiaoping made many economic changes in China.

Japan

Japan is part of Eastern Asia and consists of four main islands and thousands of small ones. The main islands, in order of size, are Honshu, Hokkaido, Kyushu and Shikoku. These islands make up more than 98 per cent of Japan.

South of Shikoku lies the Ryukyu island chain, while the Bonin Islands lie south of Tokyo. The land is mainly mountainous, with some small, fertile and thickly-populated plains around the coast. Japan lies in an unstable part of the Earth and earthquakes and volcanic eruptions are common. Mount Fuji, a dormant volcano, is Japan's highest peak. Most of the country's rivers are short and fast-flowing.

The climate varies from north to south. Hokkaido has cold, snowy winters, while Kyushu and Shikoku have mild winters and long, hot summers. The rainfall is generally abundant and forests cover about two-thirds of the land. Evergreens, such as fir and spruce, grow in the north, with deciduous trees, such as maple and oak, in the warmer centre and south. Wild animals include bears, wild boars and deer. Northern Honshu is the world's northern limit for monkeys, such as the Japanese macaque.

▲ **Spring flowers**
Each spring, the cherry trees blossom first in the warm south then in the north. This is marked by spring festivals.

Whales, dolphins and many kinds of fish live in the seas around Japan. The country is rich in waterbirds, including the cormorant, which can be trained to catch fish.

Japan lies on a part of the Earth called 'the Pacific ring of fire'. Along this zone, the Earth's plates are on the move. Under much of Japan, the Pacific plate is sinking under the Eurasian plate. Every time the Pacific plate moves downwards, it makes the ground shake. An earthquake in 1923 struck the Tokyo-Yokohama area, killing 100,000 people. Earthquakes off the coast trigger off huge waves called tsunamis. When these waves hit the shore, they do tremendous damage. The 1923 earthquake generated an 11-metre-high tsunami.

▶ **Earthquake damage**
A powerful earthquake struck Kobe in 1995, killing more than 5,000 people. Earthquakes are common in Japan.

▲ **Buddhist statue, Kyoto**
Many Japanese follow a mixture of Shintoism and Buddhism, the country's two main religions.

▲ **Warm baths**
Japanese macaques are large monkeys that live on the island of Hokkaido in the far north of Japan. In cold weather, they sometimes bathe in warm springs, known as onsen.

FACTS

JAPAN
*Nihon, or Nippon Koku –
Land of the Rising Sun*
Area: *377,801 sq km*
Population: *125,761,000*
Capital: *Tokyo*
Other cities: *Osaka, Yokohama, Nagoya*
Highest point: *Mount Fuji (3,776 m)*
Official language: *Japanese*
Currency: *Yen*

► Tokyo

Badly damaged by an earthquake in 1923 and by bombing in World War II, Tokyo is now a mainly modern city.

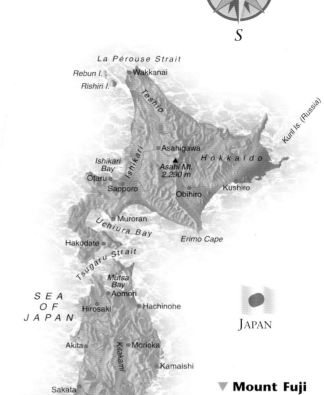

◄ Puffer fish

The poisonous puffer fish, or fugu, is found in the Pacific. Its flesh is a delicacy in Japan. To be skilled enough to cut off the poisonous parts of the fish, fugu chefs train for at least three years.

JAPAN

▼ Mount Fuji

Mount Fuji, also called Fujiyama or Fuji-san, is Japan's highest peak. Mount Fuji is a volcano which last erupted in 1708.

► Kyushu island

Kyushu is the most southerly of Japan's four main islands. Its climate is much warmer than that of northern Japan.

► Bonsai tree

Japanese gardeners have perfected the art of growing miniature trees, known as bonsai.

Japan

FOR LONG PERIODS IN THE PAST, JAPAN HAS been ruled by emperors, who held great power and, before and during World War II, its Emperor Hirohito was regarded as a god. However, after World War II, Japan became a constitutional monarchy and the emperor's role was confined to ceremonial duties. The country has an elected diet (parliament), whose members choose the prime minister.

By population, Japan ranks eighth among the countries of the world. Although the forested mountains are thinly populated, the coastal plains are among the world's most densely populated areas. About 78 per cent of the people live in cities and towns and Japan has 11 cities with more populations of more than a million.

▲ **Akashi Kaikyo Road Bridge**
Completed in 1998, the bridge linking the island of Awaji to Honshu is the world's longest cable suspension bridge.

▲ **Decorated fan**
The folding fan was probably invented in Japan around 1,300 years ago. Fans are used by entertainers.

▶ **Maglev train**
Maglev, or magnetic levitation, trains float on a fixed track without touching it. High-speed maglev trains can reach 500 kilometres per hour. Japan has an extensive rail system, especially for city commuters.

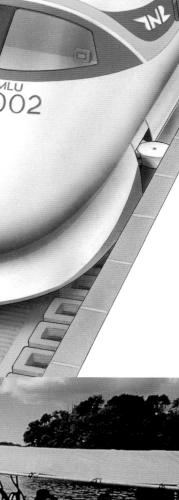

MLU 002

▲ **Whaling**
Whales are an important source of meat and oil and festivals are held to celebrate them. But international laws have forced Japanese fleets to restrict their catch.

▼ **Oyster beds**
The Japanese cultivate beds of oysters which provide delicious seafood and precious pearls.

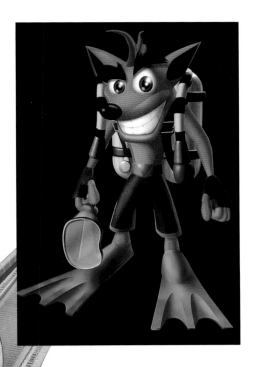

◀ **Crash Bandicoot**
Japan's top quality electronic goods, including big-name computer games are sold around the world.

▼ **Digital cameras**
Japan produces many of the world's finest cameras, lenses, video cameras and recorders, and precision instruments.

▲ **Paper lanterns**
Wood, wood pulp and paper are made in Japan. Paper products range from lanterns to sliding walls in homes.

▼ **Tea plantation**
Japan is one of the world's top ten producers of tea. Green tea is drunk at every meal.

Japan lacks natural resources and has to import food and many of the materials needed by its industries. But it ranks second only to the United States in its production of goods and services. Only 15 per cent of the land is farmed, but Japanese farmers use intensive, scientific methods to achieve some of the world's highest yields. As a result, it produces about 70 per cent of the food it needs. Rice is the leading crop, but eggs, fruit, meat, milk and vegetables are also important. Japan is the world's fourth-largest fishing nation, judged by the volume of its catch. Fish, rice and vegetables are the main ingredients for many Japanese meals, such as sushi.

The country is one of the world's great industrial and trading nations. It leads the world in producing cars, ships and steel. It is also a major producer of commercial vehicles, computers and other electronic goods, home appliances, such as refrigerators, freezers and washing machines, and radios and television sets. Japan also has a large chemical industry and silk is a traditional product. Japan imports coal and oil to fuel power stations, and it also has nuclear power stations. Many of its rivers are harnessed to produce electricity at hydroelectric plants. The country has one of the world's best transport services and one of the world's largest merchant fleets.

◀ **Sushi**
Sushi, a delicious and popular dish, consists of rice balls flavoured with vinegar and topped with raw fish, shellfish or pickles.

▶ **Robots at work**
Automation has helped Japan become the world's leading car producer. Japan also exports trucks and motorcycles. The use of modern technology keeps down the prices of Japanese goods in world markets.

Japan

▲ **Sumo wrestlers**
A form of wrestling called sumo is the leading spectator sport in Japan.

ALMOST 99 PER CENT OF THE PEOPLE OF JAPAN ARE Japanese. The chief minorities are the Ainu, most of whom live on Hokkaido, Koreans and Chinese. Some scholars think that the Ainu may have been the first people in Japan. Japanese is the official language, but it has many dialects. The Tokyo dialect is the standard form used on radio and television. Shinto and Buddhism are the chief religions and many people follow elements from both of these faiths.

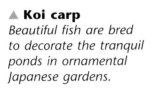

▲ **Koi carp**
Beautiful fish are bred to decorate the tranquil ponds in ornamental Japanese gardens.

Baseball and sumo (a kind of wrestling) are leading sports, together with such martial arts as judo, karate and kendo (a type of fencing). The Japanese also enjoy such Western sports as golf, skiing, tennis and volleyball. Calligraphy (beautiful handwriting), music and dance are other pastimes.

Theatre is popular. The ancient no plays are performed by actors in masks, while kabuki plays are melodramas, usually with spectacular sets. Since World War II, Japanese cinema has become popular around the world. Many films tell stories about samurai (warriors) or the shoguns (generals who ruled in the name of the emperors). Other art forms include architecture, especially Buddhist temples, literature (including poems called haiku), painting and sculpture.

▲ **Tea ceremony**
Women called geisha conduct traditional tea ceremonies. Geisha learn from childhood how to entertain guests with conversation, music and dance.

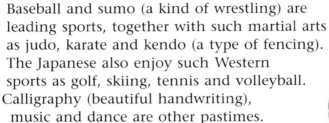

◀ **Kabuki**
In traditional Japanese kabuki plays, actors use exaggerated gestures and wear elaborate costumes.

◀ **Golden Pavilion, Kyoto**
Kyoto has many beautiful old buildings. It was Japan's capital for more than 1,000 years until 1868.

▲ **Wind-blown waves at Shichi-ri by Hiroshige**
Hiroshige (1797–1858) was one of the greatest Japanese artists of the 19th century.

◀ **Tokugawa Ieyasu**
The family of Tokugawa Ieyasu (1543–1616) ruled Japan until 1867.

◀ **Japanese samurai**
The samurai were professional knights or warriors in the Middle Ages.

▶ **Wealthy woman**
In the Tokugawa period, rich women had elaborate hair styles held together with giant pins.

Japan's modern history began when Portuguese sailors reached the islands in 1543. However, the Japanese feared that the introduction of Christianity by missionaries would lead to domination by European powers. During the 1630s, Japan cut itself off from the world and by 1640 Christianity had almost disappeared from Japan.

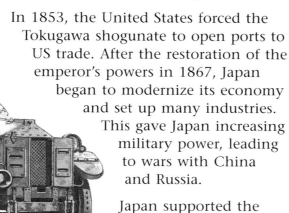

In 1853, the United States forced the Tokugawa shogunate to open ports to US trade. After the restoration of the emperor's powers in 1867, Japan began to modernize its economy and set up many industries. This gave Japan increasing military power, leading to wars with China and Russia.

Japan supported the Allies in World War I (1914–18), but in 1931 it occupied Manchuria. From 1937, it conquered much of China and during World War II (1939–45) it conquered Southeast Asia and the western Pacific. Japan was finally defeated in 1945. But, from the 1950s, it rebuilt its industries and became one of the world's leading economic powers.

▲ **Armoured car**
Japan began a war with China in 1937. This war became part of World War II (1939-45).

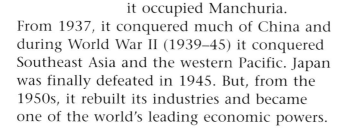

◀ **Lantern monument**
In 1945, the United States dropped atomic bombs on Hiroshima and Nagasaki. The terrible destruction led to Japan's surrender.

TIMELINE

BC	
660	*According to legend, Jimmu Tenno becomes Japan's first emperor*
AD	
300	*Yamato kingdom unifies Japan*
c.600	*Prince Shotoku makes Buddhism state religion and sets up constitution*
710	*Nara becomes Japanese capital*
785	*Heian (modern-day Kyoto) becomes capital (until 1868)*
1192	*First shogun (general) appointed*
1274	*Mongol invasion thwarted by typhoon (known as* kamikaze, *'divine wind')*
1600	*Tokugawa Period of national seclusion, called sakoku (until 1867)*
1854	*US Commander Matthew C Perry opens two Japanese ports to US trade*
1867	*Shogun rule ends*
1904	*Russo-Japanese War (until 1905): Japan defeats Russia*
1914	*World War I (until 1918): Japan sides with Germany*
1937	*Japan began a war against China*
1939	*World War II (until 1945): Japan sides with Axis Powers; US drops atomic bombs on Hiroshima and Nagasaki*
1947	*Japan adopts democratic constitution*
1989	*Emperor Hirohito dies; his son Akihito becomes emperor*

Southeast Asia

Southeast Asia consists of a large peninsula attached to India and China, together with thousands of islands to the east and south of the peninsula. Like Eastern Asia, Southeast Asia contains several fast-developing countries, notably Singapore, Thailand and Malaysia.

The peninsula contains fertile plains, where most of the people live, and some rugged mountains. The area is drained by several major rivers. The longest is the Mekong, which rises in Tibet, and flows through Laos, Thailand and Cambodia before reaching its large delta in Vietnam. Myanmar, which was called Burma until 1989, contains the Ayeyarwady (formerly Irrawaddy) River. Its delta is one of the world's great rice-growing areas.

Malaysia lies partly on the peninsula and partly on the island of Borneo. The small state of Singapore consists of about 50 islands, the Philippines contains about 7,100, and Indonesia about 13,600. The Philippines and Indonesia are largely mountainous. Both countries lie on unstable parts of the Earth and so earthquakes are common. Also, many of the highest peaks are active volcanoes. Some of the biggest volcanic eruptions in modern times have occurred in this region. They include Tambora (1815), Krakatoa (1883) and Mount Merapi (1931), all in Indonesia, and Mount Pinatubo, Philippines (1991).

Southeast Asia has high temperatures throughout the year and abundant rainfall. But while much of the peninsula has a wet and dry monsoon climate, the southern islands are rainy throughout the year. The wettest areas have evergreen rainforests which are rich in plant and animal species. Mangrove swamps also border many coasts.

▼ Komodo dragon
The Indonesian Komodo dragon is the largest living lizard.

▲ Forest ape
Orang-utans are apes found in Borneo and Sumatra. Rainforest destruction has made them rare.

▶ Mosque, Brunei
Two-thirds of the people of Brunei, a small, oil-rich country on Borneo, are Muslims. Islam is also the chief religion of Indonesia.

FACTS

BRUNEI
Negara Brunei Darussalam – State of Brunei Darussalam
Area: 5,765 sq km
Population: 290,000
Capital: Bandar Seri Begawan
Other cities: Seria, Kuala Belait
Highest point: Bukit Betalong (913 m)
Official language: Malay, English
Currency: Brunei dollar

CAMBODIA
Preah Reach Ana Pak Kampuchea – Kingdom of Cambodia
Area: 181,035 sq km
Population: 10,275,000
Capital: Phnom Penh
Other cities: Kompong Cham, Battambang
Highest point: Phnum Aôral (1,813 m)
Official language: Khmer
Currency: Riel

INDONESIA
Republik Indonesia – Republic of Indonesia
Area: 1,904.569 sq km
Population: 197,055,000
Capital: Jakarta
Other cities: Bandung, Surabaya, Medan
Highest point: Puncak Jaya (5,030 m)
Official language: Bahasa Indonesian
Currency: Indonesian rupiah

LAOS
Saathiaranarath Prachhathipatay Prachhachhon Lao – Lao People's Democratic Republic
Area: 236,800 sq km
Population: 4,726,000
Capital: Vientiane
Other cities: Savannakhet, Luang Prabang
Highest point: Mount Bia (2,817 m)
Official language: Lao
Currency: Kip

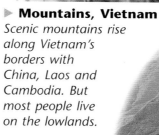

Mountains, Vietnam
Scenic mountains rise along Vietnam's borders with China, Laos and Cambodia. But most people live on the lowlands.

▲ **Crater lake**
The craters of some of Indonesia's many volcanoes are filled with mineral-rich water.

MYANMAR

LAOS

PHILIPPINES

CAMBODIA

VIETNAM

THAILAND

MALAYSIA

SINGAPORE

BRUNEI

INDONESIA

▼ **Reticulated python**
The reticulated python of Southeast Asia may grow to nine metres in length.

Map labels:

MYANMAR (BURMA) · CHINA · Mandalay · Akyab · ARAKAN YOMA · PEGU YOMA · TANEN R. · Red · Hanoi · Haiphong · Luang Prabang · Pathen (Bassein) · Yangon (Rangoon) · Chiang Mai · LAOS · Vientiane · Mawlamyine (Moulmein) · THAILAND · Nakhon Ratchasima · Da Nang · Tavoy · Bangkok · Mekong · Mergui · CAMBODIA · TONLE SAP · Nha Trang · 'Phnom Penh · VIETNAM · Gulf of Thailand · Ho Chi Minh City · Phuket · George Town · Kuala Terengganu · Ipoh · Kelang · Medan · Strait of Malacca · M A L A Y S I A · Natuna Is. · Kuala Lumpur · Johor Baharu · SINGAPORE · SARAWAK · Kapuas · Sumatra · Pontianak · Padang · Hari · BORNEO · Balikpapan · Jambi · Barito · Palembang · Belitung · Bangka · Banjarmasin · JAVA SEA · Ujung Pandang · Jakarta · Semarang · Bandung · Surabaya · Java · Malang · Bali · Lombok · Sumbawa · Flores · Ende · Sumba · Timor · Kupang · FLORES SEA · Laoag · Luzon · Mt. Pinatubo · Manila · Mindoro · PHILIPPINES · Iloilo · Panay · Tacloban · Palawan · Negros · Cebu City · Bohol · SULU SEA · Mindanao · Zamboanga · Davao · Mt. Kinabalu 4,094 m · Bandar Seri Begawan · BRUNEI · SABAH · Sandakan · Mt Apo 2,954 m · CELEBES SEA · Manado · MOLUCCA SEA · Halmahera · Moluccas · Sorong · CERAM SEA · Seram · Buru · Ambon · Palu · Sulawesi · Makassar Strait · I N D O N E S I A · Baubau · BANDA SEA · Wetar · Tanimbar Islands · Aru Is. · IRIAN JAYA · Puncak Jaya 5,030m · NEW GUINEA · Jayapura · Digul · PAPUA NEW GUINEA

▼ **Penang, Malaysia**
Penang Island lies off the northwest coast of Malaysia. During the 1800s its chief port, George Town was an important shipping centre. Today the island is a popular tourist resort.

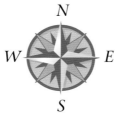

N W E S

▶ **Yangon, Myanmar**
Pagodas are towers linked by Buddhist temples. They are based on Indian stupas, which covered the remains of a king or holy man.

189

Southeast Asia

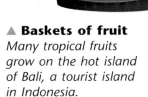

SOUTHEAST ASIA CONTAINS TEN countries. Cambodia, Malaysia and Thailand are monarchies, with elected parliaments. Brunei is also a monarchy, ruled by a sultan. Because of the country's large oil exports, the Sultan of Brunei has become one of the world's richest men.

Indonesia, the Philippines and Singapore are multi-party republics, while Laos and Vietnam are socialist republics, ruled by Communist governments. Myanmar has a military government. People who want democratic rule have been imprisoned.

Many people have also been concerned about the abuse of human rights in Indonesia. One major issue concerns East Timor. Timor is an island, consisting of two main parts. West Timor became part of Indonesia when the country became independent in 1949. But Portugal ruled East Timor until 1975, when the Portuguese left following demands by the local people for independence. Indonesian troops then occupied East Timor, which in 1976, was declared to be part of Indonesia. Since then, people in East Timor who want independence have been suppressed.

▲ **Palace statue**
Many statues stand in the Royal Palace in Bangkok, Thailand. The statues reveal the natural grace of the Thai people.

▲ **Baskets of fruit**
Many tropical fruits grow on the hot island of Bali, a tourist island in Indonesia.

▼ **Floating markets**
Canals criss-cross Bangkok, Thailand. In places, boats loaded with produce form floating markets.

▶ **Royal palace, Bangkok**
Thailand is an ancient monarchy. Unlike the other countries of Southeast Asia, it never became the colony of a Western power.

▼ **Ruby panning**
The world's finest rubies are found in river gravels in Myanmar. Thailand also produces beautiful rubies.

▼ **Oyster and pearl**
The waters off the strangely-shaped island of Sulawesi, Indonesia, are the source of oysters that yield top-quality pearls.

FACTS

MALAYSIA
Persekutuan Tanah Malaysia – Federation of Malaysia
Area: 329,758 sq km
Population: 20,565,000
Capital: Kuala Lumpur
Other cities: Ipoh, Johor Baharu
Highest point: Mount Kinabalu (4,101 m)
Official language: Malay
Currency: Ringgit

MYANMAR (BURMA)
Myanmar Naingngandaw – Union of Myanmar
Area: 676,578 sq km
Population: 45,883,000
Capital: Yangon (Rangoon)
Other cities: Mandalay, Mawlamyine (Moulmein)
Highest point: Hkakabo Razi (5,881 m)
Official language: Burmese
Currency: Kyat

PHILIPPINES
Republika ng Pilipinas – Republic of the Philippines
Area: 300,000 sq km
Population: 71,899,000
Capital: Manila
Other cities: Quezon City, Cebu, Davao
Highest point: Mount Apo (2,954 m)
Official language: Pilipino, English
Currency: Philippine peso

◀ Singapore at night
Singapore City is a bustling centre of industry, finance and trade.

About half of the people in Southeast Asia are farmers or farm workers, many of them are poor. The chief crop is rice, and Indonesia, Myanmar, the Philippines, Thailand and Vietnam rank among the world's top ten rice producers. Other major products include coffee, tropical fruits, such as pineapples, maize, rubber, tea and timber.

Brunei, Indonesia, Malaysia and Vietnam produce oil and natural gas. Other resources include coal in Indonesia and Vietnam, precious stones in Myanmar, and tin in Indonesia, Malaysia, Thailand and Vietnam.

▲ Coconut harvest
In Thailand, monkeys are trained to collect coconuts, the fruit of the coconut palm.

Nearly a third of the people live in cities in towns. The largest cities are Jakarta in Indonesia, Manila in the Philippines, Bangkok in Thailand, Ho Chi Minh City and Hanoi in Vietnam, and Singapore City. Singapore is a small county but it is the region's richest. Its highly-skilled workers have made it a major centre of manufacturing, producing chemicals, electronic products, scientific instruments, ships and textiles. Indonesia, Malaysia and Thailand have also developed industries. In contrast, neighbouring Vietnam, Cambodia, Laos are poor countries, whose economies have been badly damaged by war.

▲ Water buffalo
Farmers use water buffalo to pull ploughs in paddy fields. The animals like to wallow in mud and water.

▼ Terraced ricefields,
Step-like fields, used mainly for growing rice, have been cut into sloping land in the Philippines.

▲ Tourism
Tourism is a major industry employing many people in Thailand. Bali in Indonesia, Malaysia and the Philippines also attract many tourists.

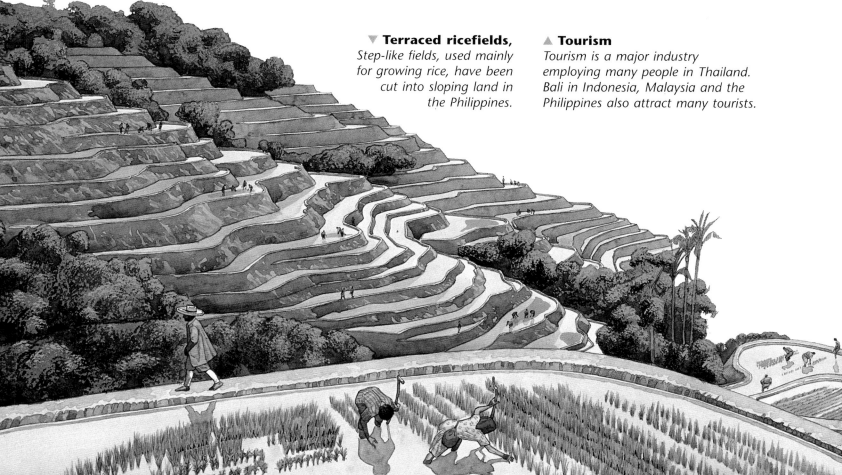

Southeast Asia

SOUTHEAST ASIA HAS ABOUT 490 MILLION PEOPLE. Most of the people are of Chinese, Indian or Malay descent. Many languages are spoken, especially in Indonesia which has about 250 languages.

Buddhism is the chief religion of much of the mainland peninsula of Southeast Asia. But Islam is important in Malaysia and Brunei. About 87 per cent of Indonesians are Muslims, although Indonesia also has Christian, Hindu and Muslim minorities. In the Philippines, about 90 per cent of the people are Christians. The Philippines is the only large country in Asia with a Christian majority. The Communist governments in Laos and Vietnam discourage any kind of religious worship.

Ancient and modern architecture is one of the region's finest art forms. Visitors to the region admire the beautiful Buddhist and Hindu temples and the Muslim mosques. Music and dance are popular, especially at festivals, such as harvest celebrations and village fairs. Indonesia has gamelan bands which play a distinctive kind of music. These bands use drums, flutes and metal gongs, together with stringed and xylophone-like instruments. Puppet dramas are popular in Indonesia and Malaysia. The person who operates the puppets tells the story and speaks the parts of the characters. The Philippines and Vietnam also have long traditions of literature.

▲ Thai children
Primary education is free in Thailand and 94 per cent of the people over 15 years old can read and write.

▲ Buddhist monks, Myanmar
About nine out of every ten people in Myanmar are Buddhists. Buddhism is also the chief religion in Cambodia, Laos and Thailand.

◄ Dancers in Bali
Many Balinese dances depict Hindu stories. Balinese orchestras called gamelan include drums, gongs, instruments like xylophones, and stringed instruments.

◄ Statue of Buddha, Vietnam
The Communist government in Vietnam discourages religious worship, but those people who are religious follow Buddha.

▲ Javanese shadow puppet
Puppeteers in Java enact a story by casting shadows of puppets on to a screen.

FACTS

SINGAPORE
Hsin-chia-p'o Kung-ho-kuo (Mandarin Chinese), Republik Singapura (Malay), Singapore Kudiyarasu (Tamil) - Republic of Singapore
Area: *618 sq km*
Population: *3.044,000*
Capital: *Singapore*
Other cities: *none*
Highest point: *Timah Hill (177 m)*
Official languages: *Chinese, Malay, Tamil, English*
Currency: *Singapore dollar*

THAILAND
Pathet Thai – Kingdom of Thailand
Area: *774,815 sq km*
Population: *60,003,000*
Capital: *Bangkok*
Other cities: *Nakhon Ratchasima, Songkhla*
Highest point: *Inthanon Mountain (2,595 m)*
Official language: *Thai*
Currency: *Baht*

VIETNAM
Công Hòa Xã Hôi Chu Nghia Viêt Nam – Socialist Republic of Vietnam
Area: *331,689 sq km*
Population: *75,355,000*
Capital: *Hanoi*
Other cities: *Ho Chi Minh City, Haiphong*
Highest point: *Fan Si Pan (3,143 m)*
Official language: *Vietnamese*
Currency: *Dong*

Longhouse
Dayak family groups in Borneo live in longhouses, which may contain up to a hundred people.

▼ Burial cave
Some people in Indonesia bury their dead in caves. Bones and skulls can often be seen. Ancestor and nature worship is practised by some tribal groups.

Leading sports include badminton, basketball and soccer, while traditional ox-races and bullfights are held at festivals in Indonesia. Thai boxing is unusual, because the boxers can use both their hands and their feet to strike their opponents.

▲ Dayak woman
Small tribal groups live in isolated parts of Southeastern Asia. They include the Dayaks of Borneo.

Many people in the cities wear Western clothes. However, many people wear sarongs, or similar items. Sarongs are long pieces of cloth which men and women wrap around their bodies. Muslim men usually wear hats, while women also cover their heads, but Muslim women seldom wear veils.

Rice is the main food in Southeast Asia. It is eaten with meat, fish and vegetables. One popular dish in Malaysia is satay. It consists of pieces of meat cooked on a skewer and dipped in a hot sauce. Adobo in the Philippines consists of chicken and pork cooked in soya sauce and vinegar.

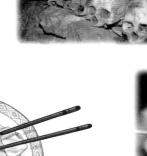

◄ Spring rolls

► Hooked on pain
Devotees undergo pain to prove their strong religious faith at the Taipsu festival, George Town, Malaysia.

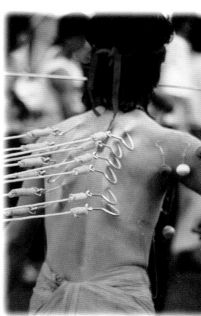

► Thai village
In the remote parts of Thailand, many people still follow ancient tribal customs.

◄ Tribal warrior
The peoples living in remote areas in the Philippines include the pygmy Negritos, descendants of people who settled there 30,000 years ago.

► Balinese procession
Processions are held on the mainly Hindu island of Bali.

193

Southeast Asia

FROM AROUND 3000BC, PEOPLE FROM CENTRAL ASIA and southern China began to settle in Southeast Asia. The original, darker-skinned peoples were forced to live in remote areas. From around AD100, a series of great kingdoms developed in the region, including the Funan kingdom in Cambodia. From the AD600s, the Srivajaya kingdom arose in Sumatra and it became a major sea power. But perhaps the best-known kingdom flourished in Cambodia between the 800s and the 1400s. This was the Angkor civilization, which was named after its capital. The ruins of its superb temples still stand today.

European influence in Southeast Asia increased after a Spanish expedition reached the Philippines in 1521. Spain gradually took control of the islands and, around 100 years later, the Dutch began to take over Indonesia, where Islam was taking over as the main religion.

In the 19th century, Britain gradually took over Singapore, Malaysia and Burma (now Myanmar), while France took over Indochina (now Cambodia, Laos and Vietnam). By the early 20th century, only Thailand remained free of foreign rule. But European power collapsed when Japan conquered Southeast Asia during World War II (1939–45). Following the defeat of Japan in 1945, the countries of Southeast Asia gradually won their independence.

▲ **Relic from Angkor**
Angkor was the ancient capital of a civilization that flourished in Cambodia between the 9th and 15th centuries. This guardian figure was found there.

▲ **Rama IV**
King Mongut (Rama IV) ruled Thailand from 1851 to 1868. He allowed many Western ideas to be introduced.

▲ **Khmer warriors**
In the 12th century, the Khmer rulers of Angkor conquered areas which are now in Laos, Thailand and Vietnam.

▼ **Angkor Wat**
The temple of Angkor Wat was the finest building in the ancient capital of Angkor. The city ruins were rediscovered in 1860.

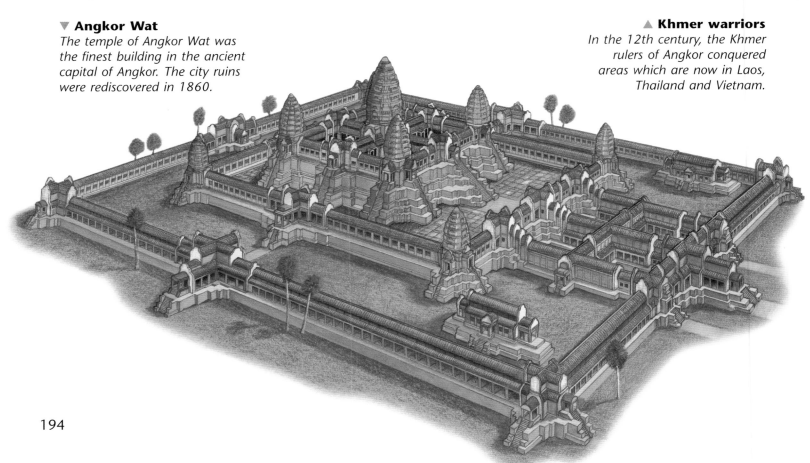

Raffles Hotel
Sir Stamford Raffles (1781–1826), the founder of Singapore City, played an important part in establishing British footholds in Southeast Asia. His name is remembered in the Raffles Hotel.

Pol Pot
Communist leader Pol Pot (c.1925–98) seized control of Cambodia in the 1970s. His forces killed thousands of Cambodians.

The ideas of Communism attracted many people in the region and Communist guerrillas fought for independence in Indochina and elsewhere. In 1954, France withdrew from Indochina, which was divided into Cambodia, Laos, North Vietnam and South Vietnam. Communist North Vietnam aided Communists in other parts of Indochina and, in the early 1960s, the United States sent troops to stop South Vietnam falling to the Communists. The war ended in 1975, when Vietnam became one nation.

The region remained disturbed, with civil war in Cambodia, and military rule in Myanmar. However, several countries made great strides in developing their economies. The region benefited from the creation of the Association of Southeast Asian Nations (ASEAN) in 1967. ASEAN was set up to help its members work together to develop their economies. By 1998, all the countries of Southeast Asia, apart from Cambodia, were members of ASEAN.

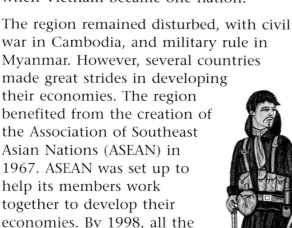

Ho Chi Minh
Communist leader Ho Chi Minh (1890–1969) led North Vietnam during the Vietnamese War (1957–75).

Vietnamese soldier

Imelda's shoes
President Marcos (1917–89) fell from power in the Philippines in 1986. He was accused of corruption. His wife, Imelda was criticized for her extravagance, after the people discovered her collection of clothes and shoes.

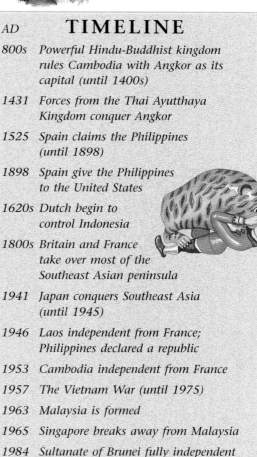

AD	**TIMELINE**
800s	Powerful Hindu-Buddhist kingdom rules Cambodia with Angkor as its capital (until 1400s)
1431	Forces from the Thai Ayutthaya Kingdom conquer Angkor
1525	Spain claims the Philippines (until 1898)
1898	Spain give the Philippines to the United States
1620s	Dutch begin to control Indonesia
1800s	Britain and France take over most of the Southeast Asian peninsula
1941	Japan conquers Southeast Asia (until 1945)
1946	Laos independent from France; Philippines declared a republic
1953	Cambodia independent from France
1957	The Vietnam War (until 1975)
1963	Malaysia is formed
1965	Singapore breaks away from Malaysia
1984	Sultanate of Brunei fully independent
1986	Revolution in the Philippines
1997	Laos and Myanmar join the Association of Southeast Asian Nations (ASEAN)

AFRICA

▲ African dog

Africa is the world's second-largest continent after Asia, covering about 20 percent of the world's land area. Most of Africa is a plateau, surrounded by narrow coastal plains. The highest peak is Kilimanjaro, a dormant (sleeping) volcano in northern Tanzania. Africa is drained by some major rivers, including the Nile, Congo and Niger.

Northwest and southwest Africa have mild Mediterranean-type climates. Between these regions lie vast, hot deserts, tropical grassland, called savanna, and dense rainforests around the Equator. The savanna is the home of such animals as antelopes, elephants, lions and zebras. Many other animals, such as gorillas and chimpanzees are found in the rainforests. The number of animals in Africa has steadily decreased because of over-hunting and because their habitats have been cleared to create farms. Still, many governments have created national parks to protect their wildlife.

The continent has two main groups of people. North of the Sahara, the Arab and Berber people speak Arabic and follow Islam. The black Africans south of the Sahara are divided into more than 1,000 ethnic groups. Some are Muslims, others are Christians, but many still follow ancient, local religions.

Africa contains many of the world's poorest countries. Many people are farmers who use simple farm tools. They produce little more than they need to feed their families and, when droughts occur, many of them starve. Agriculture employs nearly 60 percent of the people. Africa has valuable resources, such as oil and many metals, which are exported. But, apart from South Africa and Egypt, few countries have many manufacturing industries.

Around 50 years ago, European countries ruled most of the continent. Today, almost all of Africa is free. But progress has been hampered by economic problems and instability.

▲ Congo mask

▶ Painted house, Ghana

196

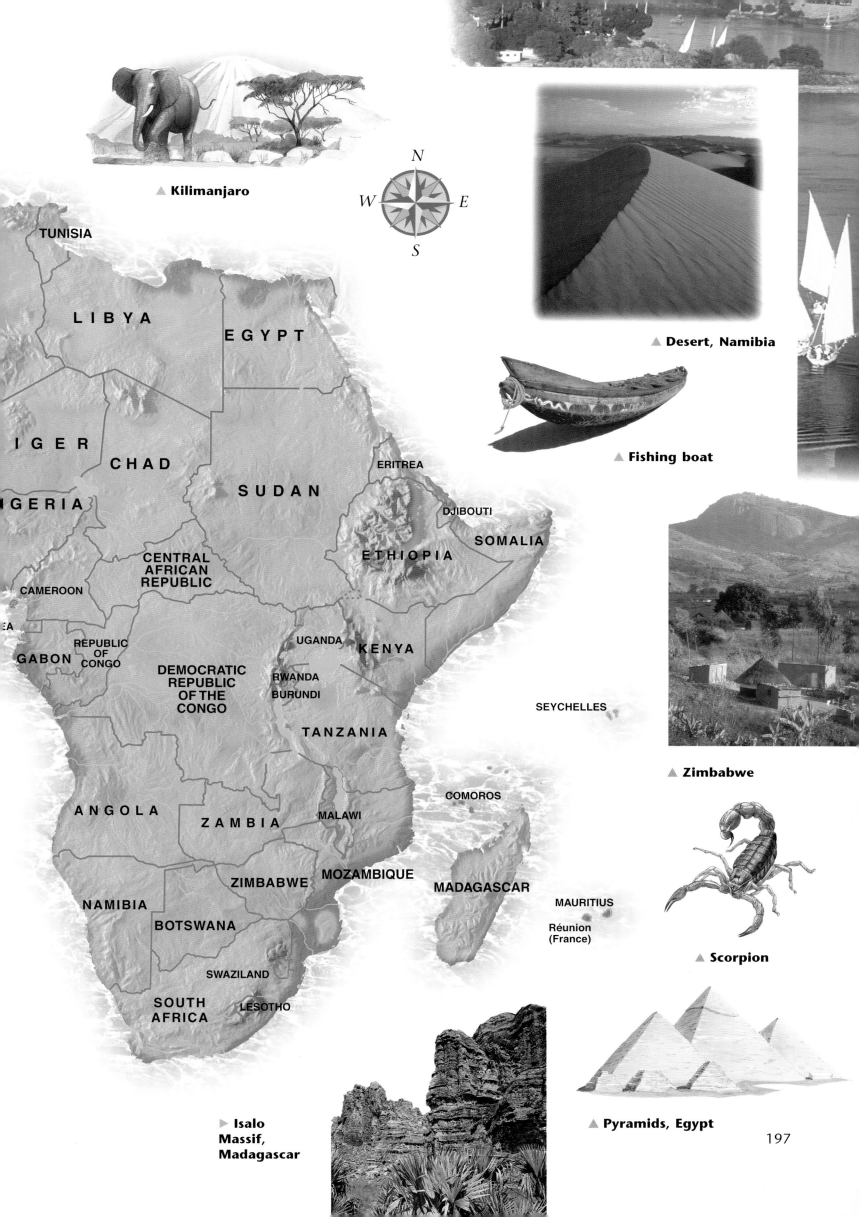

▲ Kilimanjaro

N
W E
S

▲ Desert, Namibia

▲ Fishing boat

TUNISIA

LIBYA

EGYPT

IGER

CHAD

IGERIA

SUDAN

ERITREA

DJIBOUTI

CENTRAL
AFRICAN
REPUBLIC

ETHIOPIA

SOMALIA

CAMEROON

EA

GABON

REPUBLIC
OF
CONGO

DEMOCRATIC
REPUBLIC
OF THE
CONGO

UGANDA

KENYA

RWANDA

BURUNDI

SEYCHELLES

TANZANIA

▲ Zimbabwe

ANGOLA

ZAMBIA

COMOROS

MALAWI

ZIMBABWE

MOZAMBIQUE

MADAGASCAR

NAMIBIA

BOTSWANA

MAURITIUS

Réunion
(France)

▲ Scorpion

SWAZILAND

SOUTH
AFRICA

LESOTHO

▶ Isalo
Massif,
Madagascar

▲ Pyramids, Egypt

197

North Africa

▲ **Scorpion**

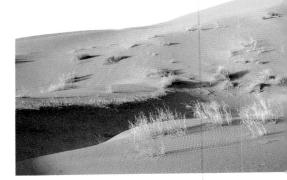

▲ **Desert, Morocco**
The Sahara is North Africa's main land region. It stretches from the Atlantic Ocean to the Red Sea.

North Africa, as shown on the map, makes up nearly half of Africa. It consists largely of a low plateau, between 150 and 600 metres above sea level. The highest peaks are in the Ethiopian Highlands in the southeast and the Atlas Mountains in the northwest. Smaller highlands made of jagged volcanic rocks rise above the plateau.

▲ **Red Sea fish**
The Red Sea is a hot, salty arm of the Indian Ocean.

Northwestern Morocco, together with the plains along the Mediterranean Sea coast, have dry, hot summers and mild, rainy winters. But the Sahara, the world's largest desert, covers more than three-fifths of North Africa. With an area of about 9,269,000 square kilometres, it is nearly as big as the United States.

The world's highest air temperature in the shade, 58°C, was recorded in the Sahara. However, nights can be cold and people wear thick cloaks to keep them warm. Several years may pass by in the Sahara with hardly any rain. Then a sudden, freak storm may cause floods. But the desert contains green oases, where water from beneath the ground reaches the surface. The region with the most water is the Nile valley. South of the Sahara, the desert merges into a dry grassland region called the Sahel. South of the Sahel lie areas of savanna, with scattered trees. Some rainforests grow in the far southeast.

The Mediterranean region contains various trees, such as oaks and olive trees. But the chief plant is the palm tree. The most important desert animal is the camel, which can travel long distances without water. Other animals include the Barbary ape in the northeast and the gelada baboon in Ethiopia.

▶ **Old volcanoes**
The Ahaggar Mountains in southern Algeria contain jagged peaks formed from the necks of ancient, extinct (dead) volcanoes. These hard rock formations are called plugs.

FACTS

ALGERIA
Jamhuriya al Jazairiya ad-Dimuqratiya ash-Shabiya – People's Democratic Republic of Algeria
Area: *2,381,741 sq km*
Population: *28,734,000*
Capital: *Algiers*
Other cities: *Oran, Constantine, Annaba*
Highest point: *Mount Tahat (2,918 m)*
Official language: *Arabic*
Currency: *Algerian dinar*

CAPE VERDE
República de Cabo Verde – Republic of Cape Verde
Area: *4,033 sq km*
Population: *389,000*
Capital: *Praia*
Other cities: *Mindelo, São Filipe*
Highest point: *Pico (2,829 m)*
Official language: *Portuguese*
Currency: *Cape Verde escudo*

CHAD
République du Tchad – Republic of Chad
Area: *1,284,000 sq km*
Population: *6,611,000*
Capital: *N'Djamena*
Other cities: *Moundou, Bongor*
Highest point: *Emi Koussi (3,415 m)*
Official languages: *Arabic, French*
Currency: *CFA franc*

DJIBOUTI
Jumhouriyya Djibouti – Republic of Djibouti
Area: *23,200 sq km*
Population: *619,000*
Capital: *Djibouti*
Other cities: *none*
Highest point: *Mousaalli (2,063 m)*
Official language: *Arabic, French*
Currency: *Djibouti franc*

▼ Waterfalls

The Atlas Mountains run though Morocco and Algeria into Tunisia. Sparkling waterfalls are common along the sheer slopes.

▲ Living by the Nile

The Nile provides much-needed water for farmers. Egypt's Nile valley is North Africa's most populated region.

MOROCCO

ALGERIA

TUNISIA

LIBYA

EGYPT

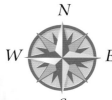
N
W   E
S

ERITREA

DJIBOUTI

Strait of Gibraltar
Madeira
Tétouan Oran
Casablanca Rabat Oujda
MOROCCO Marrakech ATLAS MTS. Constantine
CANARY Agadir Béchar Ghardaia
ISLANDS Ifni ATLAS MOUNTAINS
Las Palmas Tindouf Adrar In Salah
Western
Sahara S A H A R A
Dakhla
Cape MAURITANIA M A L I AHAGGAR MTS. AÏR
Blanc Tahat MTS.
CAPE VERDE 2,918 m
ISLANDS Nouakchott Timbuktu
Praia Kaédi Niger
Dakar N I G E R CHAD
SENEGAL Kayes
Banjul GAMBIA Ségou Niamey Zinder
Bamako Ouagadougou Kano
Kankan GUINEA BURKINA FASO Maiduguri
IVORY Tamale Zaria
COAST GHANA Kaduna Benue Yola
Garoua Sarh
CENTRAL AFRICAN
REPUBLIC

A L G E R I A
Algiers Annaba
Tunis
TUNISIA Sfax
Tripoli Misurata
Ghadames Gulf of Sirte Benghazi Darnah
L I B Y A
Alexandria Port Said
QATTARA Cairo Suez
DEPRESSION Sinai
Pen.
LIBYAN DESERT Asyût Qena
E G Y P T
Lake Aswân
TIBESTI Nasser
MTS. Nubian Port Sudan
Emi Koussi Desert
3,415 m Merowe
Faya-Largeau Atbara
Omdurman Kassala
BODÉLÉ S U D A N ERITREA
DEPRESSION El Obeid Khartoum Asmara
Lake Abéché Kosti Aksum
Chad Jabal Marrah Blue Nile ETHIOPIAN
N'Djamena 3,088 Gonder PLATEAU DJIBOUTI
Chari Lake Debre Markos Djibouti
Tana
Addis Ababa
SUDD Gore RIFT VALLEY Ogaden
White Nile ETHIOPIA Webe Shebele
Nimule SOMALIA
UGANDA KENYA

ERITREA

DJIBOUTI

CAPE VERDE
ISLANDS

MAURITANIA

MALI

NIGER

CHAD

SUDAN   ETHIOPIA

▶ Niger valley, Mali

The Niger is the third longest river in Africa. It brings water to dry areas south of the Sahara.

▶ Asni, Morocco

Asni is a market town in the foothills of the Atlas Mountains, south of the city of Marrakech.

▼ Desert landmark

The pyramids near Cairo, Egypt, were built by the rulers of ancient Egypt, around 3000BC.

199

▶ Fennec fox

North Africa

NORTH AFRICA CONTAINS 14 INDEPENDENT countries. Twelve are republics, while Morocco is ruled by a king. The government of Eritrea, which was part of Ethiopia between 1952 and 1993, is transitional – that is, its constitution has not been finalized.

The region also includes the disputed territory of Western Sahara, which was ruled by Spain until 1976. When Spain withdrew, Morocco occupied the northern two-thirds, while Mauritania took the rest. But the local Saharan people demanded independence and attacked the Moroccans and Mauritanians. Mauritania withdrew from Western Sahara in 1979, leaving Morocco in charge. But fighting continued, with the rebels claiming that their territory was an independent republic. In the 1990s, the United Nations tried to organize a vote on the future in Western Sahara.

Ethnic and religious differences have also caused many problems in North Africa. For example, in Sudan, Muslim government forces have fought civil wars with the black Africans in the south, many of whom follow traditional religions or Christianity.

▲ Peanuts
Peanuts are also called groundnuts. Sudan is a leading producer.

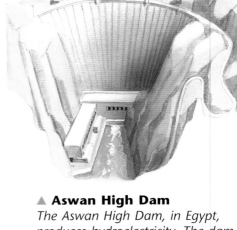

▲ **Emperors' crowns**
Axum, a historic city in northern Ethiopia, was a royal capital from 500BC. It has many ancient treasures.

▲ **Aswan High Dam**
The Aswan High Dam, in Egypt, produces hydroelectricity. The dam, which began operating in 1968, holds back Lake Nasser.

FACTS

EGYPT
*Jamhuriyat Misr al-Arabiya –
Arab Republic of Egypt*
Area: *1,001,449 sq km*
Population: *59,272,000*
Capital: *Cairo*
Other cities: *Alexandria, El Giza*
Highest point: *Jabal Katrinah
(2,637 m)*
Official language: *Arabic*
Currency: *Egyptian pound*

ERITREA
State of Eritrea
Area: *117,600 sq km*
Population: *3,698,000*
Capital: *Asmara*
Other cities: *Asseb, Keren*
Highest point: *Mount Soira
(3,103 m)*
Official language: *None*
Currency: *Nakfa*

ETHIOPIA
Federal Democratic Republic of Ethiopia
Area: *1,104,300 sq km*
Population: *58,234,000*
Capital: *Addis Ababa*
Other cities: *Dire Dawa, Harar*
Highest point: *Ras Dashen
(4,260 m)*
Official language: *None*
Currency: *Birr*

LIBYA
*Jamahiriya Al-Arabiya Al-Libiya Al
Shabiya Al-Ishtirakiya Al-Uzma – Great
Socialist People's Libyan Arab Republic*
Area: *1,759,540 sq km*
Population: *5,167,000*
Capital: *Tunis*
Other cities: *Benghazi, Misurata*
Highest point: *Bette Peak (2,286 m)*
Official language: *Arabic*
Currency: *Libyan dinar*

▲ **Market, Algeria**
On market days people buy and sell produce, meet friends and catch up on local news.

▶ **Moroccan carpets**
Highly decorated carpets are among the magnificent products made by skilled Arab and Berber craftspeople in North Africa.

▼ **Suez Canal**
The Suez Canal, which was opened in 1869, provides a short cut between the Indian Ocean and the Mediterranean Sea.

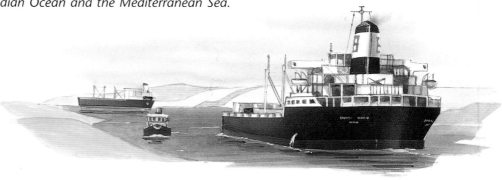

◀ Tanneries, Fez
Fez is an ancient royal capital city in Morocco. It has tanneries which prepare leather to make a wide range of leather products.

▲ Oases
In the dry desert, water is a precious resource.

Poverty is widespread in North Africa and the United Nations classifies most countries as poor, developing areas. The wealthier countries include Algeria and Libya, with their reserves of oil and natural gas, and Morocco, where phosphate rock is mined and made into fertilizer. Niger, one of the poorest countries, is a major producer of uranium.

Overall, agriculture employs more than 50 per cent of North Africans. Important crops include barley, citrus fruits and olives in the Mediterranean countries, cotton and sugar cane in the Nile valley, and dates at desert oases. South of the Sahara, millet and sorghum are leading food crops, while coffee is Ethiopia's main export.

Manufacturing is unimportant in most North African countries. The chief industrial region is the Nile delta north of Cairo, the capital of Egypt. With more than nine million people, Cairo is the largest city in North Africa. About 37 per cent of North Africans live in cities and towns. Other cities with populations of more than two million are Alexandria and El Giza, Egypt; Casablanca, Morocco; Addis Ababa, Ethiopia.

▲ Oranges, Tunisia
Oranges and other citrus fruits grow well in the sunny climate of North Africa.

◀ Decorated camel
Tourists enjoy camel rides in North Africa. The camels here have one hump. They kneel to let riders climb on and off them.

▲ Snake charmer
Snake charmers and other entertainers perform at the Djemaa El Fna, a market in central Marrakech, Morocco.

FACTS

MALI
République du Mali – Republic of Mali
Area: 1,240,192 sq km
Population: 9,999,000
Capital: Bamako
Other cities: Ségou, Mopti
Highest point: Homboro Tondo (1,155 m)
Official language: French
Currency: CFA franc

MAURITANIA
République Islamique Arabe et Africaine de Mauritanie – Islamic Republic of Mauritania
Area: 1,025,520 sq km
Population: 2,332,000
Capital: Nouakchott
Other cities: Nouadhibou, Kaédi
Highest point: Kediet Ijill (915 m)
Official language: Arabic
Currency: Ouguiya

MOROCCO
Mamlaka al Maghrebia – Kingdom of Morocco
Area: 446,500 sq km
Population: 27,020,000
Capital: Rabat
Other cities: Casablanca, Marrakech
Highest point: Jebel Toubkal (4,165 m)
Official language: Arabic
Currency: Moroccan dirham

NIGER
République du Niger – Republic of Niger
Area: 1,267,000 sq km
Population: 9,335,000
Capital: Niamey
Other cities: Zinder, Maradi
Highest point: Mount Gréboun (1,944 m)
Official language: French
Currency: CFA franc

North Africa

MOST PEOPLE IN AFRICA NORTH OF THE Sahara speak Arabic and follow Islam. As a result, the countries share a common culture, with similar kinds of buildings, literature and other art forms. However, Islamic culture has been influenced by the West in the last 100 years. Some Muslims dislike Western ideas and support groups who call for the return of fundamentalist Islamic laws and practices. In Algeria, the conflict between the government and Muslim fundamentalists has led to fighting in the 1990s.

Many languages are spoken in the countries south of the Sahara and the customs of the people vary from group to group. For example, more than 70 languages and 200 dialects are spoken in Ethiopia, which has no single official language. On the other hand, several other countries use a European language for official purposes. Soccer is the leading sport throughout the region.

Ful or *fool*, which consists of broad beans cooked in oil, is a favourite dish in Egypt and Sudan. Couscous (steamed wheat served with fish, meat, vegetables and a sauce) is popular in northwestern Africa. Ethiopians enjoy *wat*, a spicy stew eaten with a flat bread called *injera*.

▲ **Tuareg**
The Tuaregs are a nomadic tribe. Men wear clothes of a distinctive dark blue colour.

▲ **Sultan's mosque, Cairo**
The skyline of Cairo, Africa's largest city, is marked by mosques with tall minarets (prayer towers).

▲ **Egyptian hats**
Most North African men wear hats to prevent sunstroke. The hats are brimless and flat-topped.

▼ **Sand oven, Timbuktu**
Timbuktu, Mali, was once a great trading and educational city. Many ancient traditions still survive there.

FACTS

SUDAN
Jamhuryat es-Sudan – Republic of Sudan
Area: 2,505,813 sq km
Population: 27,272,000
Capital: Khartoum
Other cities: Omdurman, Khartoum North, Port Sudan
Highest point: Mount Kinyeti (3,187 m)
Official language: Arabic
Currency: Sudanese dinar

TUNISIA
Jumhuriya at Tunisiya – Republic of Tunisia
Area: 163,610 sq km
Population: 9,132,000
Capital: Tunis
Other cities: Sfax, Aryanah
Highest point: Mount Chambi (1,544 m)
Official language: Arabic
Currency: Tunisian dinar

WESTERN SAHARA
Western Sahara (disputed territory largely occupied by Morocco) – local nationalists have named it the Saharan Arab Democratic Republic
Area: 266,000 sq km
Population: 228,000
Capital: El-Aaiún
Other cities: Dakhla
Highest point: 823 m
Official language: Arabic
Currency: Moroccan dirham

▲ **Shopping day**
Bamako, capital of Mali, is a major economic centre on the Niger River.

▼ **Bedouin camp**
Arab herders called Bedouin roam North Africa's deserts in search of pasture.

◄ Sphinx, Luxor
Statues of imaginary creatures called sphinxes lined avenues to temples in ancient Egyptian cities, including Karnak at Luxor.

► Hannibal
Hannibal was born at Carthage, Tunisia. He was a brilliant general who fought the Romans.

Around 5000BC, North Africa was rainier than it is today and the Sahara was a grassland. By 4000BC the climate became drier and people moved into areas with water, notably the Nile valley. Here, from about 3100BC, Ancient Egypt developed into a major civilization. Other early cultures developed to the south in Sudan and Ethiopia, which became a Christian empire in the 4th century AD.

► Obelisk, Axum
Obelisks at Axum recall the glories of ancient Ethiopia.

In the Middle Ages, Europeans knew little about Africa south of the Sahara. But Arab and Berber traders regularly crossed the desert in camel caravans. These traders recorded several major medieval empires ruled by black Africans. They include ancient Ghana and Mali, which were based in the upper Niger River region, in what is now just Mali.

▲ Mansa Musa
Mansa Musa ruled the wealthy Mali empire from 1307. His pilgrimage to Mecca in 1324 involved a dazzling procession.

By the early 20th century, most of North Africa, apart from Ethiopia, was under European rule. However, in the 1950s and 1960s, most North African countries won their independence. Since then, they have faced many economic and political problems which have slowed down their progress.

◄ Washing day, Egypt
Egyptians in the Nile valley still use the river as they did in ancient times.

▲ Muammar Gaddafi
Colonel Gaddafi (b.1942) has led Libya's government since 1969. He has encouraged a return to the fundamental principles of Islam.

BC	**TIMELINE**
5000	*Sahara is a grassland*
4000	*Sahara begins to dry up*
3100	*Upper and Lower Egypt unite to form one of the world's greatest early civilizations*
1400	*Ancient Egypt reaches its peak*
30	*Egypt and the Mediterranean coasts of North Africa come under Roman rule*
AD	
639	*Arabs from Arabia introduce Arabic and Islam into North Africa (by 710)*
1000	*Large kingdoms arise in southwestern North Africa*
1800s	*Europeans begin to claim parts of North Africa*
1882	*British troops occupy Egypt*
1922	*Egypt becomes a partly independent monarchy*
1950s	*Independence won by Libya (1951), Sudan (1952), Morocco and Tunisia (1956)*
1960	*Chad, Mali, Mauritania and Niger become independent from France*
1962	*Algeria becomes independent*
1978	*Egypt reaches a peace agreement with Israel, after 30 years of hostility*
1991	*Algeria cancels elections won by a Muslim party; civil war begins*
1993	*Eritrea breaks away from Ethiopia after a long war*

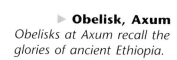

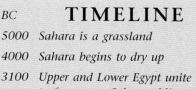

West Africa

> ▶ **Parrot**
> Parrots live in West African forests.

▲ **Merchant, Burkina Faso**
Farming employs 90 per cent of the people of Burkina Faso. About 70 per cent of the people cannot read or write.

West Africa covers about 12 per cent of Africa. It extends from Senegal and Gambia in the west to Cameroon, Central African Republic and Equatorial Guinea in the east. The largest country, Nigeria, covers about a quarter of West Africa. It contains more than half of the people of West Africa. With a population of more than 111 million, Nigeria has more people than any other African country.

The landscape consists of coastal plains and a large inland plateau. The highest peak is Mount Cameroon, an active volcano which overlooks the northern coast of Cameroon. Moist winds blow from the sea up the slopes of Mount Cameroon, bringing heavy rain. With an average of about 10,000 mm a year, the slopes of this mountain are one of the wettest places in the world.

▲ **Wide-eyed**
Bushbabies are active at night. They dart through the forest trees.

West Africa has high temperatures throughout the year and the southern coasts have rain all the year round. Dense rainforests once covered this region, but much of the original forest has been cut down to make way for farmland. The northern part of West Africa has heavy summer rains, but winters are dry. Here, the forest merges into savanna, with scattered trees, though leafy forests line the rivers.

▲ **Beautiful beaches**
The fine beaches and beautiful scenery of Ghana could attract many foreign tourists. Gambia has the most developed tourist industry in West Africa.

▲ **Waterfall, Nigeria**
Waterfalls and rapids are common features on the river Niger in Nigeria. They make long stretches of the river unnavigable.

The chief river is the Niger. It rises near the Sierra Leone-Guinea border, not far from the Atlantic Ocean. But instead of flowing directly into the sea, it follows a huge arc through Mali and Niger. Finally, it empties into the Gulf of Guinea in Nigeria. Part of Lake Chad lies in northeastern Nigeria, but the largest inland body of water entirely in West Africa is Lake Volta in Ghana. It is a reservoir, formed behind a dam.

▼ **Fishing boat, Ghana**
Lagoons line much of West Africa's coast from Liberia to Nigeria. By contrast, between Sierra Leone and Senegal, the land has sunk, flooding coastal valleys.

► Rocky outcrop
Behind West Africa's narrow coastal plains, the land rises to tablelands broken by occasional mountains.

▲ Acacia tree
Acacias are typical of tropical Africa. Many grow in the drier savanna lands of West Africa.

▲ Wild dog

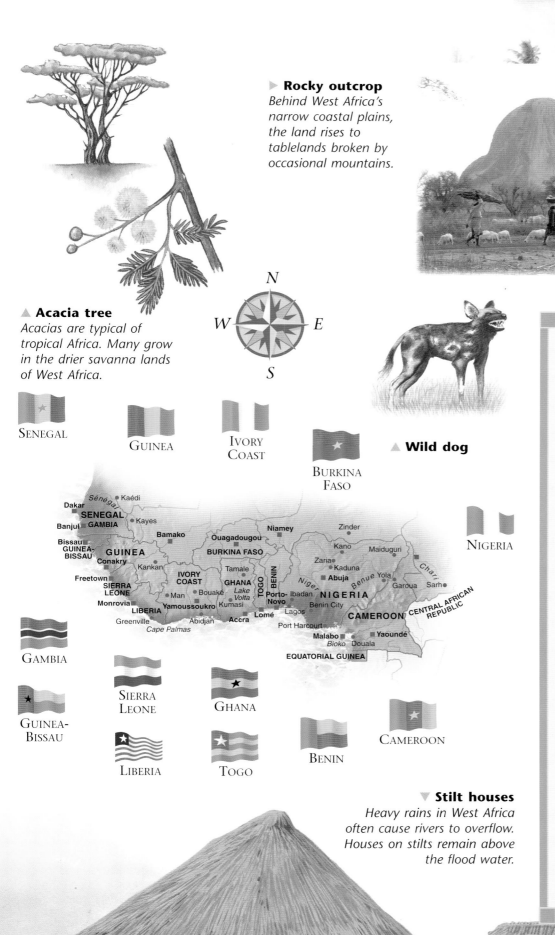

N
W · E
S

SENEGAL

GUINEA

IVORY COAST

BURKINA FASO

NIGERIA

GAMBIA

GUINEA-BISSAU

SIERRA LEONE

LIBERIA

GHANA

TOGO

BENIN

CAMEROON

Dakar · Sénégal · Kaédi
SENEGAL
Banjul · GAMBIA · Kayes
Bissau
GUINEA-BISSAU · Bamako · Ouagadougou · Niamey · Zinder
GUINEA · Conakry · Kano · Maiduguri
Kankan · BURKINA FASO · Zaria
Freetown · IVORY COAST · Tamale · Kaduna · Yola
SIERRA LEONE · Bouaké · GHANA · TOGO · Niger · Abuja · Benue · Garoua · Sarh
Monrovia · Man · Yamoussoukro · Kumasi · Porto-Novo · Ibadan · NIGERIA · Chari
LIBERIA · Lomé · Lagos · Benin City
Greenville · Abidjan · Accra · Port Harcourt · CAMEROON · CENTRAL AFRICAN REPUBLIC
Cape Palmas · Malabo · Yaoundé
Bioko · Douala
EQUATORIAL GUINEA

▼ Stilt houses
Heavy rains in West Africa often cause rivers to overflow. Houses on stilts remain above the flood water.

FACTS

BENIN
République du Bénin
Area: 112,622 sq km
Population: 5,632,000
Capital: Porto-Novo
Other cities: Cotonou, Djougou, Abomey-Calavi
Highest point: Atacora Mountains (about 610 m)
Official language: French
Currency: CFA franc

BURKINA FASO
République Démocratique du Burkina Faso – Democratic Republic of Burkina Faso
Area: 274,000 sq km
Population: 10,669,000
Capital: Ouagadougou
Other cities: Bobo-Dioulasso, Koudougou
Highest point: Aiguille de Sindou (717 m)
Official language: French
Currency: CFA franc

CAMEROON
République du Cameroun – Republic of Cameroon
Area: 475,442 sq km
Population: 13,676,000
Capital: Yaoundé
Other cities: Douala, Bafoussam, Garoua
Highest point: Mount Cameroon (4,070 m)
Official languages: French, English
Currency: CFA franc

West Africa

WEST AFRICA CONTAINS 14 COUNTRIES. MOST OF them are republics, though military groups have ruled several countries for long periods. The military leaders claimed that civilian governments were inefficient or corrupt, but they, too, were often accused of the same things.

For example, Nigeria became independent from Britain in 1960. But, since 1966, when a military group overthrew its civilian government, it has been a scene of conflict. In 1967, the Ibo people in the southeast tried to break away from Nigeria and found a separate country called Biafra, causing a civil war which ended in 1970. Military leaders again took over in 1983. Elections held in 1993 were cancelled by the military leaders, and Nigeria continued as a military dictatorship until new elections in 1999.

Military take-overs have occurred in most West African countries and two, Equatorial Guinea and Central African Republic, suffered periods of brutal military dictatorship. The only countries which have enjoyed relatively stable government are Cameroon, Ivory Coast and Senegal.

▲ **Chillies, Nigeria**
Chilli peppers are used to spice foods.

▲ **Banana seller, Ghana**
Bananas are grown throughout West Africa. But Ghana's most

▲ **Diamonds, Ghana**
Ghana produces bauxite (which is used to make aluminium), diamonds, gold and manganese. But most Ghanaians are farmers.

▲ **Salt sacks, Gambia**
Salt is obtained from seawater that is trapped in ponds and left to evaporate in the open air.

▼ **Rubber, Liberia**
Rubber trees are grown on plantations in Ivory Coast, Liberia and Nigeria. Workers tap the trees to collect a fluid called latex.

FACTS

EQUATORIAL GUINEA
República de Guinea Ecuatorial – Republic of Equatorial Guinea
Area: 28,051 sq km
Population: 410,000
Capital: Malabo
Other cities: Bata, Ela-Nguema
Highest point: Pico de Basilé (3,008 m)
Official languages: Spanish, French
Currency: CFA franc

GAMBIA
Republic of The Gambia
Area: 11,295 sq km
Population: 1,147,000
Capital: Banjul
Other cities: Serekunda, Brikama
Highest point: about 75 m
Official language: English
Currency: Dalasi

GHANA
Republic of Ghana
Area: 238,533 sq km
Population: 17,522,000
Capital: Accra
Other cities: Kumasi, Tamale, Tema, Sekondi-Takoradi
Highest point: Afadjato (885 m)
Official language: English
Currency: Cedi

GUINEA
République de Guinée – Republic of Guinea
Area: 245,857 sq km
Population: 6,759,000
Capital: Conakry
Other cities: Kankan, Nzérékoré
Highest point: Mount Nimba (1,752 m)
Official language: French
Currency: Guinean franc

◀ **Tourism, Gambia**
Tourists enjoy Gambia's beaches. They also go on organized trips, called safaris, to see wildlife.

▲ **Timber rafts**
West Africa produces valuable tropical hardwoods, such as ebony and teak.

The World Bank classifies the 14 countries of West Africa as 'low-income economies'. Agriculture employs about 55 per cent of the people, but many farmers are poor, producing little more than they need to provide for their families. In the hot, wet south, important crops include bananas, cassava, cocoa, coffee, palm oil and other palm products, rubber and yams. To the north, such crops as groundnuts, maize, millet and sorghum are important. Livestock is raised on the savanna.

Mineral resources in West Africa include oil in Nigeria and Cameroon, bauxite (aluminium ore) in Guinea, gold in Ghana, and iron ore in Liberia. Manufacturing has been growing in Ivory Coast, especially around Abidjan, around Dakar in Senegal, and in Nigeria, but elsewhere it is generally on a small scale.

About 38 per cent of the people live in cities and towns. The largest city is Lagos, former capital of Nigeria, with a population of 10,287,000. Other cities include Abidjan, Ivory Coast (2,500,000); Accra, Ghana (1,781,000); Dakar, Senegal (1,729,000); and Conakry, Guinea (1,508,000).

FACTS

GUINEA-BISSAU
República da Guiné-Bissau – Republic of Guinea-Bissau
Area: *36,125 sq km*
Population: *1,094,000*
Capital: *Bissau*
Other cities: *Bafatá, Gabú*
Highest point: *Pico de Basilé (3,008 m)*
Official language: *Portuguese*
Currency: *CFA franc*

IVORY COAST
République de la Côte d'Ivoire – Republic of the Ivory Coast
Area: *322,463 sq km*
Population: *14.347,000*
Capital: *Yamoussoukro*
Other cities: *Abidjan, Bouaké, Daloa*
Highest point: *Mount Nimba (1,752 m)*
Official language: *French*
Currency: *CFA franc*

LIBERIA
Republic of Liberia
Area: *111,369 sq km*
Population: *2,810,000*
Capital: *Monrovia*
Other cities: *Harbel, Gbarnga, Buchanan*
Highest point: *Nimba Mountains (1.380 m)*
Official language: *English*
Currency: *Liberian dollar*

NIGERIA
Federal Republic of Nigeria
Area: *923,768 sq km*
Population: *114,568,000*
Capital: *Abuja*
Other cities: *Lagos, Ibadan, Kano. Ogbomosho*
Highest point: *Dimlang Peak (2,042 m)*
Official language: *English*
Currency: *Naira*

▶ **Lagos, Nigeria**
Lagos is the largest city in West Africa. It was Nigeria's capital, but when it became congested, a new capital was built at Abuja, in central Nigeria.

◀ **Fishing, Nigeria**
Fish is an important food. Fishermen operate in inland waters, coastal lagoons and on the open sea.

207

West Africa

WITH NEARLY 210 MILLION PEOPLE, WEST AFRICA IS the most densely populated part of the continent. The people are divided into many groups. For example, Nigeria contains about 250 ethnic groups, each with its own language. Islam is the chief religion in Gambia, Guinea, Nigeria, Senegal and Sierra Leone and also in the northern parts of the other countries. But around a third of the people of West Africa are Christians and nearly a fifth follow traditional religions. Traditional religions vary, but they all contain a belief in the existence of one supreme god.

Music and dance are popular pastimes in West Africa and soccer is the leading sport. People also enjoy listening to folk tales, which storytellers pass down from one generation to the next. Cassava, maize, millet, plantains (a kind of banana) and yams are important foods. They are often eaten with spicy sauces.

▲ **African drums**
Drumming plays a major part in West African music. The main feature is the competing rhythms played by several drummers.

▲ **Yoruba, Nigeria**
The Yoruba people of Benin and southwestern Nigeria have a rich literature.

▶ **Painted homes, Ghana**
Most village homes have thick mud walls and stay cool during the day. Many are highly decorated.

FACTS

SENEGAL
République du Sénégal – Republic of Senegal
Area: 196,722 sq km
Population: 8,534,000
Capital: Dakar
Other cities: Thiès, Kaolack, Zinguinchor
Highest point: 498 m
Official language: French
Currency: CFA franc

SIERRA LEONE
Republic of Sierra Leone
Area: 71,740 sq km
Population: 4,630,000
Capital: Freetown
Other cities: Koidu-New Sembehun, Bo
Highest point: Loma Mansa (1,948 m)
Official language: English
Currency: Leone

TOGO
République Togolaise – Republic of Togo
Area: 56,785 sq km
Population: 4,230,000
Capital: Lomé
Other cities: Sokodé, Kpalimé
Highest point: Agou (986 m)
Official language: French
Currency: CFA franc

▲ **Musical instrument**
The kora is a harp-lute.

▶ **Lion coffin, Ghana**
To honour the dead, some Ghanaians buy impressive carved and elaborately painted coffins, resembling lions and other animals, or even Cadillac cars.

◀ **Nigerian family**
Many Nigerians wear traditional clothes made of brightly coloured fabrics. But nowadays many people in towns wear Western clothes.

Nok pottery, Nigeria
Pottery is still made at Nok. This village was a great artistic centre producing magnificent clay sculptures around 2,000 years ago.

Yoruba sculpture, Nigeria
Yoruba sculpture is among Africa's finest. Some sculpture was made to decorate courts. Other pieces were made for religious purposes.

The Nok culture, which is known for its beautiful terracotta sculptures, flourished in central Nigeria between 2,200 and 1,800 years ago. These sculptures are the oldest known in Africa south of the Sahara. In around AD1000, another Nigerian culture, the Ife, produced superb bronze sculptures. But the best-known bronzes were made at Benin in southern Nigeria, between 1450 and 1750. Sculpture remains important today. For example, wooden masks and carvings are made for use in traditional religious ceremonies.

Mask, Ivory Coast
Many masks were used in religious ceremonies.

More than 2,000 years ago, many cultures developed in West Africa. In the Middle Ages, northern West Africa was in contact with North Africa through the Muslim traders who crossed the Sahara on camel caravans. Europeans explored the West African coast in the mid-15th century, and soon the region became a centre of the slave trade. The Portuguese and later the Dutch, British, French, Swedes, Danes and others all traded in slaves. The British abolished the slave trade in 1807, although it continued until the mid-19th century. In the late 19th century, most of West Africa came under European rule, but independence was achieved in the 1950s and 1960s.

TIMELINE

BC	
500	Nok civilization flourishes in central Nigeria (until AD200)
AD	
1100s	Kano, in northern Nigeria, is chief trading city for Saharan camel caravans; Islam spreads through West Africa
1440s	Portuguese reach West Africa; the slave trade begins soon afterwards
1600s	Ashanti unite into a nation, which, at its peak, includes western Togo, much of Ghana and eastern Ivory Coast
1822	The American Colonization buys land in Liberia for freed slaves
1847	Liberia becomes independent
1880s	European colonization of Africa
1957	Ghana becomes independent
1958	Guinea becomes independent
1960	Benin, Burkina Faso, French Cameroon, Central African Republic, Ivory Coast, Nigeria, Senegal and Togo became independent
1961	British Cameroon becomes independent, with part joining Nigeria and part Cameroon
1965	Gambia becomes independent
1968	Equatorial Guinea and Sierra Leone become independent
1999	Civilian rule is restored in Nigeria, after a long period of military rule

Ghana's independence day
In 1957, Ghana became the first black African country to win its independence. Power passed from Britain to an elected parliament. But democracy proved fragile. Ghana, like many other African countries, suffered periods of dictatorship.

Central, Eastern and Southern Africa

▲ **Kilimanjaro**

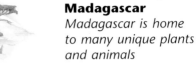

▲ **Isalo Massif, Madagascar**
Madagascar is home to many unique plants and animals

Central Africa consists mainly of the basin of Africa's second-longest river, the Congo. East Africa consists mainly of a high plateau which contains Africa's two highest peaks: Kilimanjaro in Tanzania and Mount Kenya in Kenya. Southern Africa also consists of a plateau, with some saucer-like depressions, such as the Okavango Delta in northern Botswana, where the Okavango River flows into an inland depression. The coastal plains are narrow, except in Mozambique and Somalia.

The region also includes Madagascar, the world's fourth-largest island, and the small island nations: the Comoros, Mauritius and Seychelles.

▼ **Tomato frog**

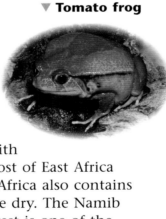

Central Africa has a hot and mostly rainy climate. Rainforests cover much of the northern part of the Congo basin, with savanna in the south. Savanna covers most of East Africa where temperatures are lower. Southern Africa also contains large areas of savanna, but some areas are dry. The Namib Desert in the southwest is one of the world's driest places, while the Kalahari in Namibia, Botswana and South Africa is a semi-desert. The southern parts of South Africa have a temperate climate. The area around Cape Town in the southwest has hot, dry summers and mild, rainy winters.

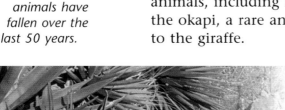

◀ **Giraffe**
Africa's savanna regions are rich in wildlife, although the numbers of animals have fallen over the last 50 years.

The Congo basin contains many rainforest animals. Some, including the mountain gorilla, found in eastern Congo and Rwanda, are now rare because of hunting. The forests also contain other endangered animals, including monkeys and the okapi, a rare animal related to the giraffe.

FACTS

ANGOLA
República de Angola – Republic of Angola
Area: 1,246,700 sq km
Population: 11,100,000
Capital: Luanda
Other cities: Huambo, Benguela, Lobito
Highest point: Môco (2,619 m)
Official language: Portuguese
Currency: Kwanza

BOTSWANA
Republic of Botswana
Area: 581,730 sq km
Population: 1,480,000
Capital: Gaborone
Other cities: Francistown, Selebi-Pikwe
Highest point: Otsa Mountain (1,489 m)
Official language: English
Currency: Pula

BURUNDI
Republika y'Uburundi, République du Burundi – Republic of Burundi
Area: 27,834 sq km
Population: 6,423,000
Capital: Bujumbura
Other cities: Gitega, Bururi
Highest point: Mount Teza (2,666 m)
Official languages: Rundi, French
Currency: Burundi franc

CENTRAL AFRICAN REPUBLIC
République Centrafricaine – Central African Republic
Area: 622,984 sq km
Population: 3,344,000
Capital: Bangui
Other cities: Berbérati, Bouar
Highest point: Mount Toussoro (1,330 m)
Official language: French
Currency: CFA franc

▲ Rural area, Zimbabwe
Southern Africa consists mainly of rolling tablelands, broken by rocky hills and mountain ranges.

▶ Chimpanzee
Chimpanzees live in central Africa's forests.

◀ Namib Desert
The Namib Desert on Namibia's coast contains huge sand dunes where few living things can survive.

▼ On the lookout
Meerkats stand upright to see any predators that may attack them.

▼ The baobab
Africa's baobab trees, recognizable by their swollen trunks, can live more than 1,000 years.

▶ Victoria Falls
One of Africa's natural wonders, the Victoria Falls lie on the Zambezi River between Zambia and Zimbabwe.

▼ Savanna stripes
Zebras are one of the most familiar animals on the African savanna. Lions prey on them.

GABON

CONGO

ANGOLA

NAMIBIA

BOTSWANA

SOUTH AFRICA

LESOTHO

CENTRAL AFRICAN REPUBLIC

RWANDA

UGANDA

BURUNDI

KENYA

SOMALIA

TANZANIA

ZAMBIA

SEYCHELLES

MAURITIUS

MALAWI

MADAGASCAR

MOZAMBIQUE

SWAZILAND

ZIMBABWE

N
W   E
S

Map labels:

CHAD

CENTRAL AFRICAN REPUBLIC
Bozoum
Bangui
Bangassou
Bomu
Uele

CAMEROON
Congo

SÃO TOMÉ & PRÍNCIPE

Libreville
REP. OF CONGO
Mbandaka
Kisangani

Cape Lopez
GABON

Cape Lopez
Brazzaville
Kinshasa
DEMOCRATIC REP. OF CONGO

Cabinda (ANGOLA)
Matadi
Kananga
Kasai
Sankuru

Luanda
ANGOLA PLATEAU
Likasi

ANGOLA
Lobito   Huambo
Lubumbashi

Namibe
Ndola
ZAMBIA
Lusaka

Cunene
Cubango
Cuito

Etosha Pan

NAMIBIA
Windhoek

Okavango Delta
Livingstone
Harare
ZIMBABWE
Bulawayo

BOTSWANA
Gaborone

KALAHARI DESERT

NAMIB DESERT

Pretoria
Maputo
Johannesburg
Mbabane
SWAZILAND
High Veld

Kimberley
Maseru
LESOTHO
Durban

SOUTH AFRICA
Great Karoo
Orange

Cape Town
East London
Cape of Good Hope
Cape Agulhas
Port Elizabeth

SUDAN

Lake Turkana
Margherita Peak 5,709 m
UGANDA
Kampala
Lake Victoria

RWANDA
Kigali
Bukavu
BURUNDI
Bujumbura

KENYA
Kisumu
Mt. Kenya ▲ 5,199 m
Nairobi
Tana

Mwanza
Dodoma
Kilimanjaro 5,895 m

ETHIOPIA
Juba

SOMALIA
Cape Caseyr
Berbera
Mogadishu
Kismayu

Mombasa
Zanzibar
Dar-es-Salaam

Lake Tanganyika
Lake Mweru
Lake Nyasa

TANZANIA
Rufiji

MALAWI
Lilongwe
Blantyre

Zambezi
MOZAMBIQUE
Beira
Limpopo

Moçambique
Mozambique Channel

INDIAN OCEAN

SEYCHELLES
Aldabra Is.
C. Delgado
COMOROS
C. d'Ambre

Antisiranana
Mahajanga
Toamasina

Antananarivo
MADAGASCAR
Fianarantsoa
C. Ste. Marie

MAURITIUS
Réunion (France)

211

Central, Eastern and Southern Africa

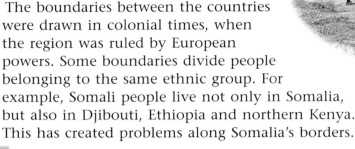

CENTRAL, EASTERN AND SOUTHERN AFRICA contain 24 independent countries and one French island territory called Réunion. The countries are all republics except for Lesotho and Swaziland, which have kings as their heads of state.

The boundaries between the countries were drawn in colonial times, when the region was ruled by European powers. Some boundaries divide people belonging to the same ethnic group. For example, Somali people live not only in Somalia, but also in Djibouti, Ethiopia and northern Kenya. This has created problems along Somalia's borders.

Many countries contain a large number of ethnic groups who have fought against each other. For example, fighting in Burundi and Rwanda between the two main groups, the Tutsi and the Hutu, has caused many deaths. Other countries which have been hit by ethnic civil wars in recent years include Angola, the Democratic Republic of Congo (formerly Zaire) and Uganda.

▲ Fishing, Luanda
Fishing is important along Angola's coast.

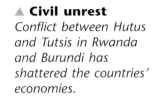

▲ Civil unrest
Conflict between Hutus and Tutsis in Rwanda and Burundi has shattered the countries' economies.

FACTS

COMOROS
République Féderale Islamique des Comores – Federal Islamic Republic of the Comoros

Area: 2,235 sq km
Population: 505,000
Capital: Moroni
Other cities: Mutsamudu
Highest point: Mount Kartala (2,361 m)
Official languages: Comorian, Arabic, French
Currency: Comorian franc

CONGO, DEMOCRATIC REPUBLIC OF
République Démocratique du Congo – Democratic Republic of the Congo

Area: 2,344,858 sq km
Population: 45,234,000
Capital: Kinshasa
Other cities: Lubumbashi, Mbuji-Mayi, Kisangani
Highest point: Mount Ruwenzori (5,109 m)
Official language: French
Currency: Congolese franc

CONGO, REPUBLIC OF
République du Congo – Republic of Congo

Area: 342,000 sq km
Population: 2,705,000
Capital: Brazzaville
Other cities: Pointe-Noire, Loubomo
Highest point: Létéki (1,040 m)
Official language: French
Currency: CFA franc

GABON
République Gabonaise – Gabonese Republic

Area: 267,668 sq km
Population: 1,125,000
Capital: Libreville
Other cities: Port-Gentil, Franceville
Highest point: Mount Iboundji (1,190 m)
Official language: French
Currency: CFA franc

▼ South African apples

▶ Mombasa, Kenya
Sunshine and stunning beaches attract tourists to tropical Africa. Camel rides and boat trips are extra attractions.

▼ Cape Town, South Africa
The flat-topped Table Mountain overlooks Cape Town, South Africa's legislative capital. Cape Town is also a major port and industrial city.

Gold mine
Gauteng province, in South Africa, is famous for its gold mines.

South Africa was once ruled by people of European origin. They introduced a policy called apartheid, under which white people ruled the country, while non-whites had little or no power. This led to a civil war. But, in 1994, a government with a black majority took power. Nelson Mandela (*b.*1918), the leading black opponent of apartheid, became president.

Agriculture employs about two-thirds of the people in this part of Africa. However, many farmers produce barely enough to support their families. Crops vary with the climate. Maize is a major food crop, while coffee, tobacco, tea and various fruits are important exports.

Natural resources include oil in Gabon and Congo, copper in Zambia, diamonds in Botswana, the Democratic Republic of Congo, South Africa and Namibia, gold in South Africa, and uranium in Namibia and South Africa. However, apart from South Africa, the region lacks large-scale manufacturing centres. South Africa is the most developed country in this part of Africa, but many black Africans are poor.

▲ **Ostrich chicks, South Africa**
Ostriches were once prized for their feathers, but today farmers rear them mainly for their hides.

▼ **Hydroelectric dam, Zimbabwe**
Hydroelectric plants at dams on Africa's great rivers produce large amounts of cheap electricity.

Large cities include Kinshasa, Democratic Republic of Congo (population, 3,804,000); Cape Town, South Africa (2,350.000); Luanda, Angola (2,250,000); Nairobi, Kenya (2,000,000); and Maputo, Mozambique (2,000,000).

FACTS

KENYA
Jamhuri ya Kenya – Republic of Kenya
Area: 580,367 sq km
Population: 27,364,000
Capital: Nairobi
Other cities: Mombasa, Kisumu, Nakuru
Highest point: Mount Kenya (5,199 m)
Official languages: Swahili, English
Currency: Kenya shilling

LESOTHO
Kingdom of Lesotho
Area: 30,355 sq km
Population: 2,023,000
Capital: Maseru
Other cities: Maputsoe, Teyateyaneng
Highest point: Thabana Ntlenyana (3,482 m)
Official languages: Sotho, English
Currency: Loti

MADAGASCAR
Repobikan'i Madagasikara, République de Madagascar – Republic of Madagascar
Area: 587,041 sq km
Population: 13,705,000
Capital: Antananarivo
Other cities: Toamasina, Antsirabe, Mahajanga
Highest point: Maromotro (2,876 m)
Official languages: Malagasy, French, English
Currency: Malagasy franc

MALAWI
Dkiko la Malawi – Republic of Malawi
Area: 118,484 sq km
Population: 10,016,000
Capital: Lilongwe
Other cities: Blantyre, Mzuzu
Highest point: Sapitwe (3,000 m)
Official language: English
Currency: Malawi kwacha

MAURITIUS
Republic of Mauritius
Area: 2,040 sq km
Population: 1,134,000
Capital: Port Louis
Other cities: Beau Bassin-Rose Hill, Vacoas-Phoenix
Highest point: Piton de la Rivière Noire (826 m)
Official language: English
Currency: Mauritian rupee

MOZAMBIQUE
República de Moçambique – Republic of Mozambique
Area: 801,590 sq km
Population: 18,028,000
Capital: Maputo
Other cities: Beira, Nampula
Highest point: Mount Binga (2.436 m)
Official language: Portuguese
Currency: Metical

Central, Eastern and Southern Africa

MOST PEOPLE IN CENTRAL, EASTERN, AND SOUTHERN Africa speak a language which belongs to the Bantu family. For example, Kongo in Central Africa, Swahili, in East Africa, and Xhosa and Zulu in South Africa are all Bantu languages.

The ancestors of the Bantu-speaking people were farmers who used iron tools. They originated in an area around the Cameroon-Nigeria border. More than 2,000 years ago they began to spread east and south, taking over the land from earlier people, including pygmies, who were forest hunters, and the Khoikhoi and San (also called Hottentots and Bushmen) in the southwest. A few of these people survive, mainly in remote areas. Today, however, the main minority groups in this region are people of European and Asian origin.

Christianity was introduced to the region by European missionaries and today, more than three-fifths of the people are Christians. About 22 per cent of the people still follow ancient religions, most of which include the worship of ancestors and spirits. However, they all contain a belief in one supreme god. Muslims, who make up 12 per cent of the population, are found mostly in Eastern Africa.

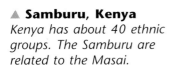

▲ **Samburu, Kenya**
Kenya has about 40 ethnic groups. The Samburu are related to the Masai.

FACTS

NAMIBIA
Republic of Namibia
Area: *824,292 sq km*
Population: *1,584,000*
Capital: *Windhoek*
Other cities: *Swakopmund, Rundu*
Highest point: *Brandberg (2,580 m)*
Official language: *English*
Currency: *Namibian dollar*

RWANDA
Republika y'u Rwanda,
République Rwandaise –
Republic of Rwanda
Area: *26,338 sq km*
Population: *6,727,000*
Capital: *Kigali*
Other cities: *Ruhengeri, Butare*
Highest point: *Mount Karisimbi*
(4,507 m)
Official languages: *Kinyarwanda, French*
Currency: *Rwanda franc*

SAO TOME & PRINCIPE
República Democrática de São Tomé e
Príncipe – Republic of São Tomé
and Príncipe
Area: *964 sq km*
Population: *135,000*
Capital: *São Tomé*
Other cities: *Trinidade, Santana*
Highest point: *Pico de São Tomé*
(2,024 m)
Official language: *Portuguese*
Currency: *Dobra*

▲ **Hutu woman**
Rwanda and Burundi have three ethnic groups: the majority Hutu, the Tutsis and a few pygmies.

▼ **Cathedral, Gabon**
Christianity is the leading religion in Central, Eastern and Southern Africa. It was introduced by European missionaries, such as the explorer David Livingstone.

▼ **African township, near Cape Town**
Many people in South Africa live in townships near the cities. Poor housing, poverty and crime are major problems.

◄ Mbuti dancer
The Mbuti are a pygmy people who live in the northeastern part of the Democratic Republic of Congo.

▼ Xhosa village
Most Xhosa live in villages in southeastern South Africa. Many depend on money sent from relatives in the cities.

Music, especially drumming, is a major traditional art form while, in recent years, several kinds of African jazz have developed. Dance is another popular pastime. Soccer is the leading sport, while other European games, including cricket and rugby football, are widely played in South Africa, Zimbabwe and some other countries.

The carving of wooden statues and masks is another art form, especially in Central Africa. Many beautiful carvings are made for use in traditional religious ceremonies and then they are thrown away. Pottery, baskets decorated with beads and shells, and dance shields are other items of great beauty made throughout the region.

▲ Congo mask
Masks are used in ceremonies by people who follow traditional religions.

Most black Africans are poor and the most common dishes are simple. For example, maize is the principal food in savanna regions. It is pounded into a flour which is cooked to form a kind of porridge. This porridge is eaten with beans or some other vegetable, often cooked in a spicy sauce. Fish and meat are eaten when people can afford them.

► Zulu women
Many Zulu men work in mines and industries. Their wives stay at home and bring up the children.

◄ Kenyan athlete
East Africa has produced many fine athletes. They excel at long-distance running, especially at high altitudes.

FACTS

SEYCHELLES
Repiblik Sesel,
République des Seychelles –
Republic of the Seychelles
Area: 455 sq km
Population: 77,000
Capital: Victoria
Other cities: none
Highest point: Morne Seychellois
(905 m)
Official languages: Creole, French,
English
Currency: Seychelles rupee

SOMALIA
Jamhuriyadda Dimugradiga ee
Soomaaliya – Somali Democratic
Republic
Area: 637,657 sq km
Population: 9,805,000
Capital: Mogadishu
Other cities: Hargeisa, Kismayu,
Berbera
Highest point: Surud Ad (2,408 m)
Official languages: Somali, Arabic
Currency: Somali shilling

SOUTH AFRICA
Republic of South Africa
Area: 1,221,037 sq km
Population: 37,643,000
Capital: Pretoria (executive),
Cape Town (legislative),
Bloemfontein (judicial)
Other cities: Johannesburg, Durban,
Port Elizabeth
Highest point: Mount aux Sources
(3,282 m)
Official languages: Afrikaans,
English, Ndebele, North Sotho,
South Sotho, Swazi, Tsonga,
Tswana, Venda, Xhosa, Zulu
Currency: Rand

SWAZILAND
Umboso weSwatini –
Kingdom of Swaziland
Area: 17,364 sq km
Population: 926,000
Capital: Mbabane
Other cities: Manzini, Nhlangano
Highest point: Mount Emlembe
(1,862 m)
Official languages: Swazi, English
Currency: Lilangeni

TANZANIA
Jamhuri ya Muungano wa Tanzania –
United Republic of Tanzania
Area: 883,749 sq km
Population: 30,494,000
Capital: Dodoma
Other cities: Dar es Salaam,
Mwanza, Tanga
Highest point: Kilimanjaro
(5,895 m)
Official languages: Swahili, English
Currency: Tanzania shilling

Central, Eastern and Southern Africa

SCIENTISTS HAVE FOUND FOSSILS OF HUMAN-LIKE creatures which lived in East Africa over two million years ago. Some experts think that East Africa was the place where the first modern human beings appeared.

In ancient times, people who lived by hunting animals and gathering plant foods lived in Central, Eastern and Southern Africa. However, from about 2,000 years ago, Bantu-speaking farmers entered the region from the northwest. They intermarried with the original people, though some small groups of hunter-gatherers survived in remote areas.

▲ Australopithecus
Remains of early humans have been found in Eastern and Southern Africa.

▲ Great Zimbabwe
The unusual stone buildings in Zimbabwe were built by a Shona kingdom.

The Bantu-speaking people developed large and powerful kingdoms. However, from the 15th century, most of these kingdoms were weakened by the slave trade. Europeans seldom ventured inland and, instead, they traded with Africans on the coast. The coastal Africans attacked other groups living inland, capturing slaves whom they sold to the European traders. Large areas were devastated by these wars.

One important European settlement was made by the Dutch at Cape Town in 1652. Cape Town, which served Dutch ships sailing between Europe and southeastern Asia, became the nucleus of what is now South Africa.

▼ Bartolomeu Dias memorial
Dias (1450–1500) was the first European to reach the southern tip of Africa.

▼ Zulu wars
The Zulus of South Africa clashed with white settlers, called Boers, in 1836. Zulu resistance to European rule continued until 1879, when British troops finally defeated the Zulu army.

FACTS

UGANDA
Republic of Uganda
Area: 241,038 sq km
Population: 19,741,000
Capital: Kampala
Other cities: Jinja, Mbale, Masaka
Highest point: Mount Ruwenzori (5,109 m)
Official language: English
Currency: Uganda shilling

ZAMBIA
Republic of Zambia
Area: 752,618 sq km
Population: 9,215,000
Capital: Lusaka
Other cities: Ndola, Kitwe, Mufulira
Highest point: 2,067 m
Official language: English
Currency: Zambian kwacha

ZIMBABWE
Republic of Zimbabwe
Area: 390,757 sq km
Population: 11,248,000
Capital: Harare
Other cities: Bulawayo, Chitungwiza, Mutare
Highest point: Mount Inyangani (2,593 m)
Official language: English
Currency: Zimbabwe dollar

REUNION (FRANCE)
Département de la Réunion – Department of Réunion
Area: 2,510 sq km
Population: 664,000
Capital: Saint-Denis
Other cities: Le Port, Le Tampon
Highest point: Piton des Neiges (3,069 m)
Official language: French
Currency: French franc

◄ **Fort Jesus, Kenya**
Fort Jesus, a Portuguese fortress built in the 1590s, helped the Portuguese to win control over the East African coast.

Explorers mapped the African interior during the 19th century. Towards the end of the 19th century, European countries divided the region into colonies. Most colonies won their independence in the 1960s and 1970s, although Zimbabwe (formerly called Rhodesia) did not get majority rule until 1980. Another country, Namibia, which had been ruled by South Africa, became independent only in 1990.

In 1948, South Africa began a policy known as apartheid. Through this policy, whites controlled the government and the economy. Blacks had no right to vote and their movements and actions were strictly controlled. From the 1960s, most other countries condemned South Africa's policies, while opposition from black groups in South Africa gradually increased.

▲ **The Great Trek**
In 1836, Boer (Afrikaner) settlers moved inland from the Cape to escape British rule.

Apartheid was ended in the early 1990s and, in 1994, all the people of South Africa voted in elections which resulted in a black majority government. However, the new South Africa faces huge problems in trying to tackle the great poverty suffered by most of its black people.

TIMELINE

AD

1100	Arab traders settle East Africa
1250	Great Zimbabwe develops
1487	Bartolomeu Dias rounds the southern tip of Africa
1652	Dutch found Cape Town
1800s	European colonization
1958	Central African Republic becomes independent
1960	Democratic Republic of Congo, Republic of Congo, Gabon and Somalia become independent
1962	Burundi, Rwanda and Uganda become independent
1963	Kenya becomes independent
1964	Lesotho, Malawi, Tanzania and Zambia become independent
1965	Zimbabwe becomes independent
1966	Botswana becomes independent
1968	Swaziland becomes independent
1975	Angola and Mozambique become independent
1988	South Africa grants independence to Namibia
1994	South Africa's first democratic elections

▲ **Racial discrimination**
From 1948, life in South Africa was dominated by racial restrictions. Whites and non-whites were forced to use separate basic facilities and even to swim on separate beaches.

▶ **Nelson Mandela**
Nelson Mandela led the African National Congress to victory in South Africa's first non-racial elections in 1994. Mandela had been imprisoned from 1962 to 1990 for his opposition to apartheid.

217

OCEANIA

The continent of Oceania is also known as Australasia. It takes in a vast swathe of the Pacific Ocean, stretching from the Tropic of Cancer in the north to Stewart Island (New Zealand) in the south.

Australia takes up 90 percent of the land area. Too large to be called an island, it is normally rated as the smallest of the world's major landmasses. Australia is a land of empty deserts, ringed by temperate grassland and rainforest, scrub and tropical rainforest.

To its north lies the island of New Guinea. The western part of the island (Irian Jaya) is ruled by the Asian nation of Indonesia, while the eastern part forms the independent state of Papua New Guinea. To the southeast lies the island chain of New Zealand, dominated by North and South Island. These are beautiful, cooler lands with volcanic mountains, deep sea inlets, forests and grassy plains.

Strung out across the open Pacific Ocean are thousands of tiny coral and volcanic islands, grouped into nations and dependent territories. They are sometimes divided into ethnic regions. Micronesia includes Guam, Kiribati (pronounced Kiribass), Mariana, Marshall and Caroline Islands. Melanesia includes Papua New Guinea, Vanuatu, New Caledonia, Fiji and the Solomon Islands. Polynesia includes Tonga, Samoa, the Line Islands, Tuvalu, French Polynesia and Pitcairn Island.

▲ **Maori meeting house**

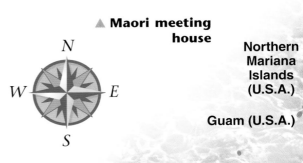

Northern Mariana Islands (U.S.A.)

Guam (U.S.A.)

PALAU   Car

INDONESIA

M E

Irian Jaya

PAP NE GUIN

AUSTRALIA

◄ **Charlotte Sound**

◄ **Kiwi, New Zealand**

▼ **Uluru (Ayers Rock), Australia**

▲ Southern Alps, New Zealand

▶ Beach houses, Fiji

Wake Island (U.S.A.)

MARSHALL ISLANDS

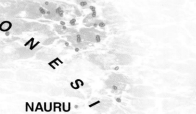

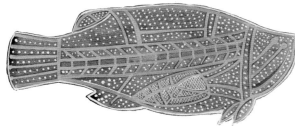

C R O N E S I A

nds

NAURU

SOLOMON ISLANDS

TUVALU

Line Islands

K I R I B A T I

N
E
S
I
A

VANUATU

WESTERN SAMOA

FIJI

American Samoa (U.S.A)

Cook Islands (N.Z.)

French Polynesia (France)

▲ Aboriginal art

▲ Kiwi fruit

TONGA

NEW CALEDONIA (France)

Pitcairn (U.K.)

Kermadec Is. (N.Z.)

Norfolk Is. (Australia)

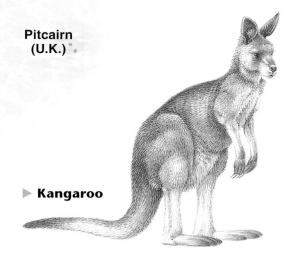

▶ Kangaroo

NEW ZEALAND

Chatham Islands (N.Z.)

▶ Sydney Opera House

▶ Tuatara

219

Australia

▲ **Thorny devil**

Australia lies between the Indian Ocean and the Pacific. Its northern shores are washed by warm, tropical seas while the island of Tasmania to the south lies in cooler, stormier waters. Its northeastern coast is protected by the world's longest coral reef, the Great Barrier, which extends for 2,027 kilometres from north to south.

The east coast is paralleled by the mountains of the Great Dividing Range, whose southern section is called the Australian Alps. Australia's longest river, the Murray, whose chief tributary is the Darling, rises near Mount Kosciusko.

The interior of Australia is mostly wilderness, made up of the Great Victoria Desert, the Gibson and the Great Sandy Desert. There are rocky outcrops, salt pans, dried-out lakes and, along the Great Australian Bight, the flatlands of the Nullarbor Plain. Surrounding the deserts are regions of dry scrub, savanna grassland, tropical rainforest, temperate grassland and eucalyptus forest. Australia's habitats are home to many species of animal and bird found nowhere else on Earth, such as koalas, kangaroos, Tasmanian devils and duck-billed platypuses.

▲ **Great Barrier Reef**
Divers explore this magical world of corals, brilliantly-coloured fish, giant clams and starfish.

▲ **Jenolan Caves**
Stalactites and stalagmites have been formed by water dripping through these limestone caves in New South Wales.

The great majority of Australians live in the big cities of the eastern, southern and southwestern coasts, where the temperature is pleasant and the ground is fertile. Where possible, the interior, or 'outback', is occupied by huge cattle and sheep stations or by mines.

▲ **Murray River turtle**

◄ **Phillip Island**
Collapsing rock has formed stacks and pinnacles off Phillip Island, near Melbourne. Bordered by Western Port and the Bass Strait, the island is famous for its fairy penguins.

FACTS

AUSTRALIA
Commonwealth of Australia
Area: *7,682,300 sq km*
Population: *18.7 million*
Capital: *Canberra*
Other cities: *Sydney, Melbourne, Brisbane, Adelaide, Perth*
Highest point: *Mt Kosciusko (2,229 m)*
Official language: *English*
Currency: *Australian dollar*

Lake Eyre

This vast salt lake lies below sea level in South Australia. It is normally dry, as the river waters that feed it soon evaporate, leaving the ground encrusted with mineral salts.

▲ **The Olgas**

Known to the Aborigines as Kata Tjuta (meaning 'many heads'), these strange, rounded domes of rock rise 50 kilometres to the west of Uluru (Ayers Rock).

AUSTRALIA

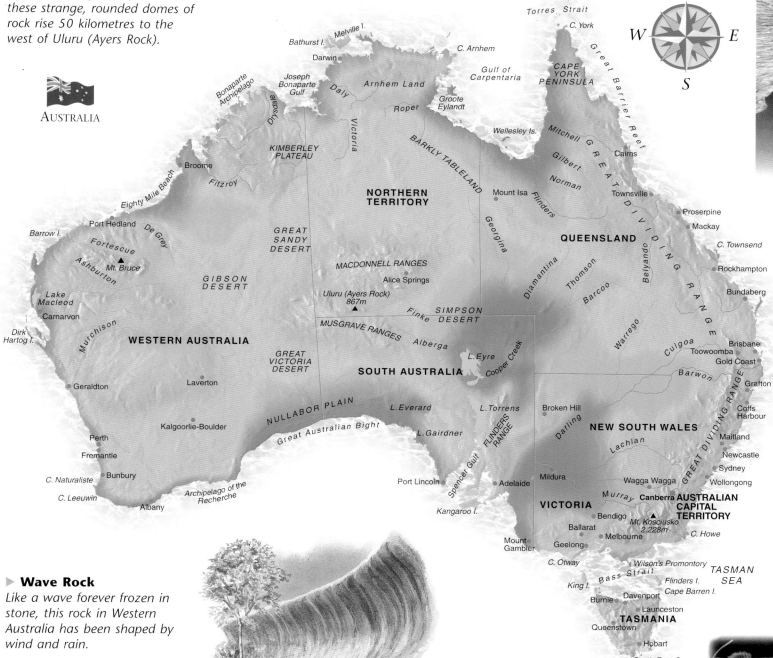

Torres Strait
C. York
Melville I.
Bathurst I.
C. Arnhem
Darwin
Gulf of Carpentaria
CAPE YORK PENINSULA
Bonaparte Archipelago
Joseph Bonaparte Gulf
Daly
Arnhem Land
Roper
Groote Eylandt
Great Barrier Reef
Drysdale
Victoria
Wellesley Is.
Mitchell
Gilbert
Norman
Cairns
KIMBERLEY PLATEAU
Broome
Fitzroy
BARKLY TABLELAND
Mount Isa
Flinders
Townsville
GREAT DIVIDING RANGE
Eighty Mile Beach
NORTHERN TERRITORY
QUEENSLAND
Proserpine
Mackay
Port Hedland
De Grey
Barrow I.
Fortescue
GREAT SANDY DESERT
C. Townsend
Ashburton
Mt. Bruce
GIBSON DESERT
MACDONNELL RANGES
Alice Springs
Georgina
Diamantina
Thomson
Belyando
Barcoo
Rockhampton
Lake Macleod
Uluru (Ayers Rock) 867m
SIMPSON DESERT
Bundaberg
Carnarvon
Murchison
Finke
MUSGRAVE RANGES
Alberga
Warrego
Culgoa
Brisbane
Dirk Hartog I.
WESTERN AUSTRALIA
GREAT VICTORIA DESERT
L. Eyre
Cooper Creek
Barwon
Toowoomba
Gold Coast
Geraldton
Laverton
SOUTH AUSTRALIA
Grafton
NULLABOR PLAIN
L. Everard
L. Torrens
Broken Hill
Darling
NEW SOUTH WALES
Coffs Harbour
Perth
Kalgoorlie-Boulder
Great Australian Bight
L. Gairdner
FLINDERS RANGE
Lachlan
Maitland
Fremantle
Newcastle
C. Naturaliste
Bunbury
Spencer Gulf
Port Lincoln
Adelaide
Mildura
Wagga Wagga
Sydney
Wollongong
C. Leeuwin
Albany
Archipelago of the Recherche
Kangaroo I.
Murray
Canberra
AUSTRALIAN CAPITAL TERRITORY
VICTORIA
Bendigo
Mt. Kosciusko 2,228m
C. Howe
Ballarat
Melbourne
Mount Gambier
Geelong
C. Otway
Wilson's Promontory
TASMAN SEA
King I.
Bass Strait
Flinders I.
Cape Barren I.
Burnie
Davenport
Launceston
TASMANIA
Queenstown
Hobart
South East C.

N W E S

Wave Rock

Like a wave forever frozen in stone, this rock in Western Australia has been shaped by wind and rain.

Byron Bay

Unspoilt sands and rolling breakers extend for huge distances along the northern coast of New South Wales.

Up a gum tree

The furry, grey koala spends most of its time in eucalyptus trees. Babies remain in a pouch on their mother's body for six months after birth.

Australia

AUSTRALIA IS AN INDEPENDENT NATION organized on federal lines. Canberra is its purpose-built centre of government, located in Australian Capital Territory (ACT) in the southeast of the country. The most populous states are New South Wales and Victoria in the southeast, followed by Queensland in the northwest, Western Australia, South Australia and the island of Tasmania. The wilderness of Northern Territory is very sparsely populated.

▲ **A birds-eye view**
A helicopter carries tourists over blue seas and reefs off Cairns, in northern Queensland. Australia is becoming a major world tourist destination.

▲ **Harbour Bridge**
This Sydney landmark, built in 1932, links the north and the south of the city.

Australia also governs some outlying ocean territories, including Norfolk Island, Christmas Island, Cocos (Keeling) Islands, the Ashmore and Cartier Islands, the Coral Sea Islands, Heard Island and Macdonald Island.

Originally a group of British colonies, the Commonwealth of Australia still has the British monarch as its head of state. However there is widespread political support for the country becoming a republic. Although cultural links with the British Isles are still strong, Australia now sees itself as a Pacific nation and part of the Pacific regional economy.

▼ **Australian wines**
Vineyards were planted in South Australia's Barossa Valley in the 1840s. Quality has steadily improved and today Australia produces a wide variety of excellent wines, which are exported around the world.

▲ **Christmas Island**
A territory of Australia, tiny Christmas Island lies in the Indian Ocean, about 300 kilometres south of Java. Its population is mostly made up of Malays and Chinese, who mine phosphates or work in tourism.

▼ **Road train**
Without a traffic jam in sight, thundering long-distance trucks kick up clouds of dust as they haul multiple trailers across the wilderness.

Sheep shearing
Part of Australian folklore, the sheep shearer was once the mainstay of the Australian economy. He is still kept busy today.

Pineapple plantation
Queensland's hot, moist climate produces tropical crops such as pineapples, bananas and sugarcane.

Cattle and sheep were traditional strengths of the Australian economy. Ranches, known as 'stations', may still occupy over 10,000 square kilometres and light aircraft may be needed to cross a single farm. Agriculture produces wheat, sugarcane, tropical and temperate fruit. Australian wines are now famous around the world.

The gold rush of the 1850s opened up Australia to mining, and today the country is a major exporter of iron ore, nickel, bauxite, gold, tin, uranium, zinc and tungsten. There are reserves of coal, oil and timber and the rivers of the Snowy Mountains, part of the Great Dividing Range, provide hydroelectric power.

Wind power
A windmill stands as a landmark in the empty outback. In these dry and dusty lands, wind may be used to raise precious underground reserves of water, for storage in tanks.

Manufactured goods include textiles, chemicals and vehicles. Australia has become a major centre of television, film and newspaper publishing.

It has also become a popular tourist destination, with people flying in to enjoy sun and surf, watersports, wilderness trekking and lively city entertainments.

Kangaroos
There are over 50 species of Australian kangaroo, large and small, grey and reddish brown, living in forests, grasslands and coastal regions. All leap on immensely powerful back legs.

Camels in the desert
Dromedaries (single-humped camels) were first brought to Australia in the 1860s. They were used by the explorer Robert O'Hara Burke (1820–61) and may still be seen in the deserts today.

Wildfire!
When the bush is as dry as tinder, one stray spark can set off a blaze. Vast bushfires are all too common, threatening property and life.

Australia

PEOPLE DESCENDED FROM THE FIRST Australians are known as Aborigines. Various Aboriginal cultures and languages exist in different parts of Australia. The Torres Strait islanders of the far north form another ancient, related ethnic group.

Experts at desert survival, the Aboriginal peoples experienced persecution and violence in the years following the arrival of the British in the 1770s. They were excluded from citizenship until 1967. Today Aborigines number only about one percent of the population. They face many economic and political problems, and are lodging claims to possession of their ancient lands. There has been a great revival of Australian and international interest in their culture and traditions in recent years.

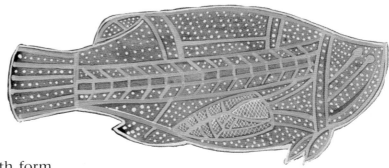

▲ **Aboriginal art**
Traditionally, the Aborigines used natural pigments for their paintings on rock faces and on panels of bark. Many paintings featured the animals that the people hunted for food.

▲ **Citizens**
Aborigines may be greatly outnumbered in their own land, but at least they now face the future as full citizens.

▲ **Artist at work**
Recent years have seen great international interest in contemporary Aboriginal art. Today, painters use modern paints and materials, but the abstract patterns they produce often refer back to ancient, spiritual beliefs.

▲ **In remembrance...**
The National War Memorial is linked by Anzac Parade to the new parliament building in Canberra. Anzac Day (April 25th) commemorates members of the Australia and New Zealand Army Corps killed in World War I (1914–18).

▼ **Barbie time**
The barbecue has become a famous Australian tradition – and nothing can beat freshly-caught seafood grilled in the open air

▶ New Australians
Since 1974, many immigrants have arrived in Australia from Southeast Asia. This Vietnamese mother and child live in the west Melbourne suburb of Footscray.

▼ Aussie football
Australian Rules football is a tough, fast game with 18 players per side.

▲ In the surf
These days, Australian life centres on the beach as much as the outback. Swimmers, lifeguards and surfers like to show off their muscles to the crowds.

The great majority of Australians are of English, Scottish, Irish or Welsh descent. The English language is spoken throughout the country, but with its own accent and many local phrases and expressions. However in the last 20 years, Australia has become much more of a multi-ethnic society. It has also been settled by Scandinavians, Poles, Dutch, Germans, Italians, Greeks, Lebanese, Indians, Chinese, Thais and Vietnamese.

Many Australians enjoy outdoor activities and sports, such as tennis, sailing, swimming, surfing, cricket, rugby and Australian rules football. Cities such as Sydney and Melbourne have become centres for the arts in recent years. Whilst 85 percent of Australians live in the modern coastal cities or in other towns, there is still a widespread fascination with the folklore and traditions of the outback. Life in the more remote areas can be tough. If there is an accident at a remote sheep station, doctors may have to fly in by plane. A teacher may have to give lessons to his or her pupils from hundreds of kilometres away, over a two-way radio link.

▲ Sydney Opera House
Completed in 1973, this inspiring building rises from Sydney harbour like a series of billowing white sails.

▲ The flying doctor
Australia pioneered airborne medical services. The Flying Doctor Service provides medical advice by radio and aircraft for emergencies on remote farms.

◀ Exploring Australia
More and more city-based Australians are beginning to explore their own vast back country, with its extraordinary landscapes.

225

Australia

HUMANS MAY HAVE FIRST ENTERED AUSTRALIA over 50,000 years ago, when low ocean levels made it easier to cross from Southeast Asia. Over the ages, waves of Aboriginal peoples moved across the land, following river valleys, hunting, fishing and trading.

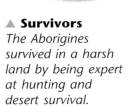

Portuguese, Spanish and Dutch explorers sailed into Australian waters, but it was the British who came to stay. In 1770 Captain James Cook (1728–79) landed at Botany Bay, and explored the east coast, which he named New South Wales.

In 1787 the British founded a prison colony at Sydney Cove. By 1823 New South Wales had become a full colony, and explorers began to set out into the harsh lands of the interior. During the 1850s, the tranportation of prisoners ended and free settlers from the British Isles arrived in their thousands, lured on by the discovery of gold. New colonies were founded in Western and South Australia, and Victoria and Tasmania broke away from New South Wales. Many Aborigine populations were killed or treated with violence. In Tasmania, all the original inhabitants were slaughtered by British settlers.

▲ Survivors
The Aborigines survived in a harsh land by being expert at hunting and desert survival.

▲ Rock art
To the ancient Aborigines, the rocks of Australia were part of a sacred landscape. Rock faces were carved, and later painted, with magical markings.

▼ The convicts
Transported across the world, often for petty crimes, British prisoners arrived to a regime of brutal punishment.

▶ Victorian age
As part of the British empire, Australia developed rapidly during the reign of Queen Victoria (1838–1901).

▲ Ned Kelly (1855–80)
Outlaw, bushranger and bank robber, the notorious Ned Kelly was hunted by the police, before a final shoot-out wearing home-made iron armour. Kelly was hanged in Melbourne jail.

▲ The rush for gold
In the 1850s, gold was discovered in New South Wales and Victoria. Prospectors arrived from around the world, hoping to make their fortune.

▼ Voices to be heard
In the last 20 years Australia's Aborigines, with their flag of black, yellow and red, have been vocal in their demand for social justice and land rights. They have forced many white Australians to reconsider their country's history.

The separate colonies federated as the Commonwealth of Australia in 1901. During the century that followed, Australia fought as part of the British Empire in two world wars, in 1914–18 and again in 1939–45, when the country was threatened by the Japanese invasion of Southeast Asia. In the 1950s, there was a new wave of settlement from Britain.

After the 1960s, Australia gradually moved away from its close political and commercial links with Britain. Japan and the United States became major trading partners. At the same time, racist laws came to an end, allowing settlement by non-European ethnic groups and civil rights for the Aboriginal population. Australia was changing rapidly.

BC	TIMELINE
50,000	Possible first Aboriginal settlement
12,000	New waves of settlement
AD	
1606	Dutch discover Australia
1644	Abel Tasman maps Australian coast
1770	James Cook claims New South Wales for Britain
1788	Prison colony founded at Sydney
1825	Tasmania becomes separate colony
1829	Western Australia colonized
1836	South Australia colonized
1840	Convict transportation ends (by 1868)
1851	Australian Gold Rush (until 1861)
1851	Victoria becomes separate colony
1855	Self-government (by 1856)
1859	Queensland becomes new colony
1901	Commonwealth of Australia founded
1914	World War I (until 1918): many Australian troops die at Gallipoli
1927	Federal government moves to Canberra
1931	Statute of Westminster: Australia confirmed as independent
1939	World War II (until 1945)
1950s	Large-scale immigration
1967	Aborigines recognized as citizens
1974	Racist imigration policies abandoned
1992	Mabo court ruling confirms Aboriginal right to claim ancestral lands

New Zealand
and the Pacific islands

▼ **Tuatara**

The Pacific Ocean is surrounded by a 'ring of fire', a region of intense volcanic activity. Erupting volcanoes under the ocean have thrown up chains of small islands. In the warm waters of the South Pacific, many of these have become rimmed with growths of coral. Sometimes the volcanic rock has then collapsed, leaving a coral ring called an atoll.

▲ **Clouds of steam**
There are seven spectacular geysers at Whakarewarewa, near Rotorua, on North Island, New Zealand.

New Zealand lies 1,920 kilometres to the southeast of Australia. It is home to many unusual animals, including the kiwi, a flightless bird. Its two main parts, North Island and South Island, are separated by the Cook Strait. North Island has rich farmland, active volcanoes, geysers and hot springs. Its major cities include Wellington, the capital, and Auckland. South Island has the snowy peaks of the Southern Alps, the fertile Canterbury Plains and the cities of Christchurch and Dunedin.

Papua New Guinea is another volcanic country, with remote, forested mountains and lush, tropical valleys, a habitat for the world's biggest butterflies. Mount Wilhelm is the highest point in Oceania. National territory also includes many smaller islands, including the Bismarck archipelago and the northern Solomons.

The nations and territories of the open Pacific are made up of thousands of tiny, scattered islands. Some are low-lying coral reefs and atolls, while others are mountainous, with tropical forest.

◀ **Navilu Island, Fiji**
Fiji is made up of 800 or small tropical islands, many of them uninhabited.

FACTS

AMERICAN SAMOA
Territory of American Samoa
UNINCORPORATED US TERRITORY

Land area: *200 sq km*
Population: *0.03 million*
Capital: *Pago Pago*
Official languages: *Samoan, English*
Currency: *US dollar*

FIJI
Republic of Fiji

Land area: *18.333 sq km*
Population: *0.8 million*
Capital: *Suva*
Other towns: *Lautoka, Nadi, Ba, Labasa*
Highest point: *Mt Tomanivi (1.323m)*
Official language: *English*
Currency: *Fiji dollar*

FRENCH POLYNESIA
Térritoire de la Polynésie Française
OVERSEAS TERRITORY OF FRANCE

Land area: *3,270 sq km*
Population: *0.2 million*
Capital: *Papeete*
Official languages: *French, Tahitian*
Currency: *CFP franc*

KIRIBATI
Republic of Kiribati

Land area: *717sq km*
Population: *0.08 million*
Capital: *Bairiki*
Highest point: *Banaba summit (81m)*
Official language: *English*
Currency: *Australian dollar*

MARSHALL ISLANDS
Republic of the Marshall Islands

Land area: *181 sq km*
Population: *0.1 million*
Capital: *Dalap-Uliga-Darrit*
Other town: *Ebeye*
Highest point: *Nowhere more than 6m*
Official languages: *English, Marshallese*
Currency: *US dollar*

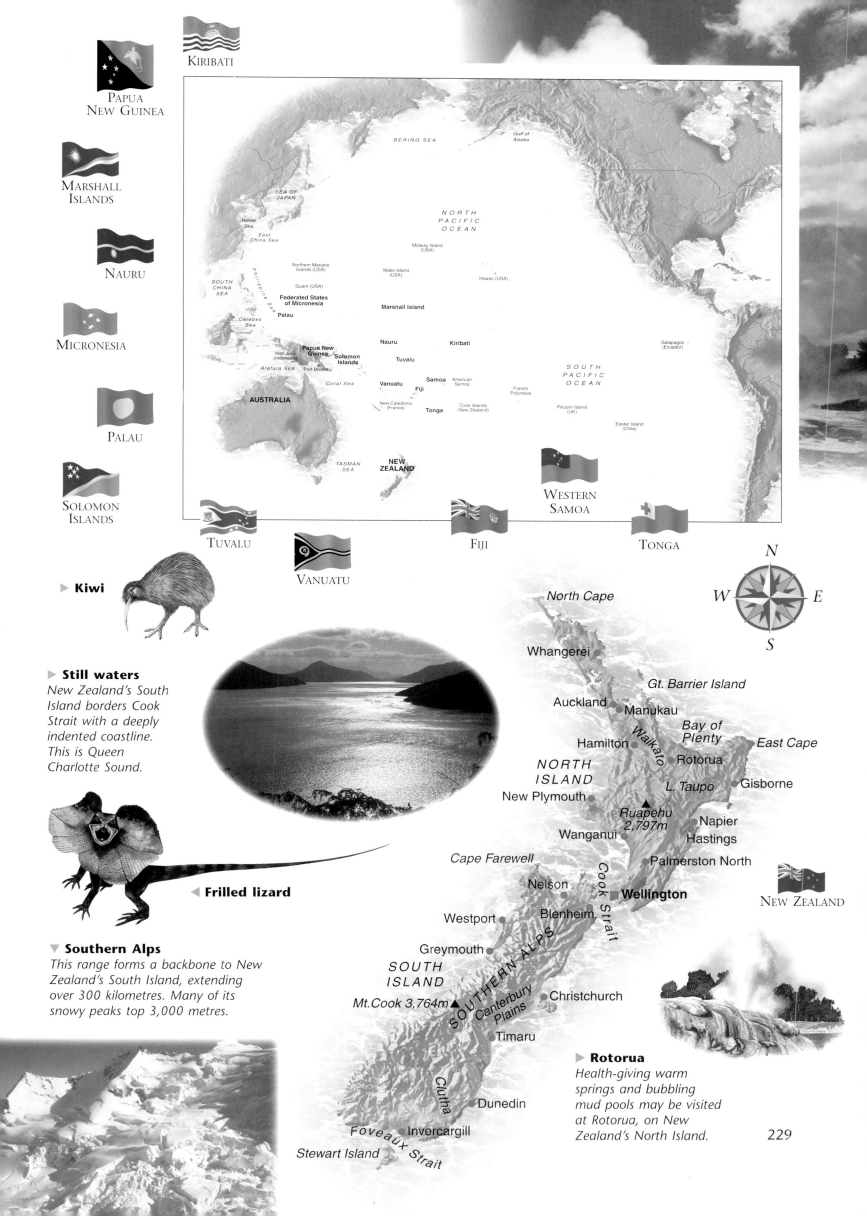

KIRIBATI

PAPUA
NEW GUINEA

MARSHALL
ISLANDS

NAURU

MICRONESIA

PALAU

SOLOMON
ISLANDS

TUVALU

VANUATU

FIJI

WESTERN
SAMOA

TONGA

BERING SEA

Gulf of
Alaska

SEA OF
JAPAN

Yellow
Sea

East
China Sea

NORTH
PACIFIC
OCEAN

Midway Island
(USA)

Wake Island
(USA)

Hawaii (USA)

Northern Mariana
Islands (USA)

Guam (USA)

SOUTH
CHINA
SEA

Federated States
of Micronesia

Palau

Marshall Island

Celebes
Sea

Nauru

Kiribati

Galapagos
(Ecuador)

Irian Jaya
(Indonesia)

Papua New
Guinea

Solomon
Islands

Tuvalu

SOUTH
PACIFIC
OCEAN

Arafura Sea

Port Moresby

Coral Sea

Vanuatu

Fiji

Samoa

American
Samoa

French
Polynesia

AUSTRALIA

New Caledonia
(France)

Tonga

Cook Islands
(New Zealand)

Pitcairn Island
(UK)

Easter Island
(Chile)

TASMAN
SEA

NEW
ZEALAND

NEW ZEALAND

▶ Kiwi

▶ Still waters
New Zealand's South
Island borders Cook
Strait with a deeply
indented coastline.
This is Queen
Charlotte Sound.

◀ Frilled lizard

▼ Southern Alps
This range forms a backbone to New
Zealand's South Island, extending
over 300 kilometres. Many of its
snowy peaks top 3,000 metres.

N
W   E
S

North Cape

Whangerei

Gt. Barrier Island

Auckland   Manukau

Bay of
Plenty

East Cape

Hamilton   Waikato   Rotorua

NORTH
ISLAND

L. Taupo   Gisborne

New Plymouth

Ruapehu
2,797m

Napier

Wanganui   Hastings

Cape Farewell   Palmerston North

Nelson   Cook Strait   Wellington

Westport   Blenheim

Greymouth   SOUTHERN ALPS

SOUTH
ISLAND

Mt.Cook 3,764m   Canterbury
Plains   Christchurch

Timaru

▶ Rotorua
Health-giving warm
springs and bubbling
mud pools may be visited
at Rotorua, on New
Zealand's North Island.

Clutha

Dunedin

Foveaux   Invercargill

Stewart Island   Strait

229

New Zealand and the Pacific islands

NEW ZEALAND IS AN INDEPENDENT NATION WHICH has the British monarch as its head of state, reflecting its historical links with the British empire. It also rules over two self-governing Pacific territories, Niue and the Cook Islands.

New Zealand has become one of the most important economies on the Pacific 'rim'. It raises large numbers of sheep and cattle and exports wool, meat (especially lamb) and dairy products. It also produces grain, vegetables and fruit, including apples and kiwi fruit. Power is generated from hydroelectric schemes and also from geothermal plants, which convert the heat from undergound volcanic activity into electricity. Tourism is a fast-growing industry.

Papua New Guinea is another independent democracy which is formally headed by the British monarch. The plantations of Papua New Guinea produce coffee, tea, rubber, palm-oil and copra (dried coconut). Villagers also clear forest to grow crops such as sweet potato, maize and bananas for their own needs. The country has rich mineral reserves, including gold, silver and copper.

▲ Kiwi fruit
The juicy green Chinese gooseberry, marketed under a name more associated with New Zealand, has become a major export.

▲ Northern skyline
Auckland is New Zealand's largest city. Founded in 1840, it is an important seaport and industrial centre.

◄ Shearing time
The New Zealand climate and terrain is ideal for raising sheep. Meat and wool are exported all over the world.

▼ In Papua New Guinea
Local markets may sell pigs, poultry, yams, taro (a root vegetable), sago (cereal from the sago palm), bananas or sweet potatoes.

FACTS

MICRONESIA
Federated States of Micronesia
Land area: *702 sq km*
Population: *0.1 million*
Capital: *Kolonia*
Other towns: *Wenu, Lelu*
Highest point: *Totolom (719 m)*
Official language: *English*
Currency: *US dollar*

NAURU
Republic of Nauru, Naoero
Land area: *21 sq km*
Population: *0.01 million*
Seat of government: *Yaren*
Highest point: *Western Nauru (70 m)*
Official languages: *Nauruan*
Currency: *Australian dollar*

NEW CALEDONIA
Nouvelle Calédonie et Dépendances
OVERSEAS TERRITORY OF FRANCE
Land area: *18,580 sq km*
Population: *0.2 million*
Capital: *Nouméa*
Official language: *French*
Currency: *CFP franc*

The smaller islands of the Pacific include independent monarchies and republics as well as overseas territories such as New Caledonia and French Polynesia. Scattered locations, small populations and lack of fertile land offer limited economic opportunities for the islanders. Many grow coconuts and tropical fruits, raise pigs and chickens and catch fish. Local dishes are often made with breadfruit, cassava or sweet potatoes. Sugarcane is grown as a cash crop in Fiji, and cocoa in the Solomon Islands. Islands such as Tahiti in French Polynesia have developed a tourist industry. Nickel is mined in New Caledonia. Tiny Nauru has been mined into a desert for its phosphates, which are used in the manufacture of fertilizers. Phosphate supplies in Kiribati have already been exhausted.

◀ **Sand and sun**
Vanua Levu and its surrounding islands, in northern Fiji, attract tourists to their palm-fringed, tropical beaches.

▲ **Pearl fishers**
The warm waters of the South Pacific are ideal for oysters and for pearls. Pearls develop when a small grain of sand enters the oyster shell.

◀ **Harvesting coconuts**
Dried coconut flesh, or copra, is produced on many Pacific islands. The hairy brown coconuts are the seeds of large green fruits, which grow at the top of the palm.

▼ **Wedding party, Tonga**
Guests gather for a royal wedding in Tonga. The chiefdoms of this Polynesian island group were united as a single kingdom in 1845.

FACTS

NEW ZEALAND
Dominion of New Zealand
Area: 268,680 sq km
Population: 3.8 million
Capital: Wellington
Other cities: Auckland, Christchurch, Dunedin
Highest point: Mt Cook (3,754 m)
Official language: English
Currency: NZ dollar

PALAU
Belu'u Era Balau
Land area: 508 sq km
Population: 0.02 million
Capital: Koror
Other towns: Melekeiok, Garusuun, Malakal
Highest point: (217 m)
Official languages: English, Palauan
Currency: US dollar

New Zealand and the Pacific islands

THE FIRST INHABITANTS OF THE SOUTH PACIFIC WERE THE ancestors of today's Melanesian, Micronesian and Polynesian peoples. Over several thousand years, they explored the world's largest ocean in their canoes, settling land from the Hawaiian islands in the north to Easter Island in the east, where they built large, mysterious stone statues. Their last great expansion took place over a thousand years ago, when a Polynesian people called the Maoris settled New Zealand.

The tattooed Maoris were fierce warriors and hunters of gigantic flightless birds called moas, now extinct. Today Maoris make up about nine percent of New Zealand's population and have preserved a keen sense of their history and culture.

▲ **Maoris**

▲ Tree carving
Carvings, customs and spirtual beliefs link many widely scattered Pacific islands, revealing common ancestry.

▼ Fiji traditions
Fiji was settled by both Melanesians and Polynesians, and later by Indians.

▲ A Melanesian potter
Melanesian peoples are found across the South Pacific, from Papua New Guinea to Fiji.

▶ Maori crafts
Intricate, swirling designs are a part of the Maori tradition, carved in wood, whale ivory or stone.

▼ The warrior age
The Maoris defended their lands with war canoes. Fortifications called pa *were surrounded by ditches and fences of stakes called palisades.*

▼ Meeting-house
The wharerunanga, *with its elaborately carved roof and doorposts, is a Maori meeting-house built in the traditional style.*

◀ **European encounter**
In 1769 the English explorer Captain James Cook sailed around New Zealand, charting the waters and meeting tattooed Maori warriors.

FACTS

SOLOMON ISLANDS
Land area: 27,600 sq km
Population: 0.4 million
Capital: Honiara
Other towns: Gizo, Kieta, Auki
Highest point: Mt Popomanaseu (2,331 m)
Official language: English
Currency: Solomon Islands dollar

TONGA
Kingdom of Tonga;
Pule'anga Fakatu'i'o Tonga

Land area: 750 sq km
Population: 0.1 million
Capital: Nukualofa
Other towns: Neiafu, Pangai
Highest point: 1,046 m
Official languages: Tongan, English
Currency: pa'anga

TUVALU
Southwest Pacific State of Tuvalu
Land area: 24 sq km
Population: 0.01 million
Capital: Fongafale
Other towns: Vaitupu, Niutao
Highest point: Nowhere more than 6 m
Official languages: English, Tuvaluan
Currency: Australian dollar

VANUATU
Ripablik blong Vanuatu
Land area: 14,800 sq km
Population: 0.2 million
Capital: Port-Vila
Other town: Luganville
Highest point: Tabwémasana (1,879 m)
Official languages: Bislama, English, French
Currency: vatu

WESTERN SAMOA
Independent State of Western Samoa
Land area: 2,830 sq km
Population: 0.02 million
Capital: Apia
Other towns: Lalomanu, Falevai, Tuasivi, Falealupo
Highest point: Mt Sisili (1,859 m)
Official languages: Samoan, English
Currency: tala

The Dutch explored New Zealand coasts in 1642, which were visited again by the English navigator Captain James Cook (1728–79), in 1769. In the following century, many British settlers came to New Zealand, attracted by the moderate climate, green pastures and discoveries of gold. They encountered warlike resistance from the Maoris. By 1852 New Zealand was a self-governing British colony and in 1907 it became a Dominion within the British Empire (having a similar status to Canada). New Zealand fought in two world wars (1914–18 and 1939–45) and maintained close ties with Britain. However by the 1970s, New Zealand was seeing itself more as a Pacific nation, expanding its trade with Japan, the USA and Australia. Today's New Zealanders enjoy a good standard of living and an outdoor life, sailing, skiing and playing team sports such as rugby and cricket.

◀ **The All Blacks**
In rugby football, New Zealand's national team has regularly been a world beater. Games start with a Maori chant and dance, the haka.

▼ **Going to church**
The Tongans had converted to the Christian faith by 1860, and today Protestant churches are well attended.

New Zealand and the Pacific islands

THE SMALLER PACIFIC ISLANDS WERE ALSO TAKEN over by colonists in the 1800s, mostly Germans, French, British and Americans. Some of the Europeans came to live on the islands, but most merely acted as colonial governors. Workers were brought to Fiji from India and their descendants became a large part of that island's population. Some colonies passed from British to Australian or New Zealand rule, but by the 1980s large areas of the Pacific were ruled by small, independent states.

▲ **House of spirits**
Most people in Papua New Guinea are Christians, but traditional beliefs in spirits and magic are still very widespread.

Ancient customs on the Pacific islands changed greatly during the colonial period, as Christian missionaries banned old forms of worship and people found work on plantations or in mines.

The last land in the region to be opened up to the outside world was Papua New Guinea. Its forest and mountain communities were so isolated from each other, that as many as 800 different languages had grown up there. The tribal costumes worn by many of the communities may still be seen at festivals today. They include spectacular feathers, bone ornaments and body paint.

The way of life in the Pacific today is a mixture of the traditional and the modern. Social life is still marked by feasting, dancing and singing. However modern communications have shrunk the great distances between one group of islands and another, and transport is by light aircraft and motor-boat as well as by canoe.

▲ **Faces of Papua**
Papua New Guinea is home to hundreds of different ethnic groups. Many of them wear elaborate face and body decorations.

▲ **Ocean rider**
This canoe is of Samoan design. The Samoan Islands were colonized by Lapita seafarers, ancestors of today's Polynesians, as early as 1000BC.

▶ **Trobriand islander**
The people of the Trobriand or Kiriwina Islands are taller and lighter-skinned than most other Melanesians.

AD	TIMELINE
c.900	Polynesians (Maoris) settle New Zealand
c.1200	Rise of Polynesian chiefdoms
1526	Portuguese discover Papua New Guinea
1642	Tasman discovers New Zealand
1769	James Cook explores New Zealand coast (until 1777)
1815	First British settlers in New Zealand
1840	Treaty of Waitangi: New Zealand linked to Australia
1842	Formation of French Polynesia
1845	Maori uprising (until 1847)
1851	New Zealand becomes a separate British colony
1860	Maori uprisings (until 1872)
1884	British and Germans claim New Guinea
1907	New Zealand self-governing
1914	World War I (until 1918): Australia occupies German New Guinea
1939	World War II (until 1945)
1970	Tonga becomes independent
1975	Papua New Guinea and Tuvalu become independent
1978	Solomon Islands become independent
1980	Vanuatu becomes independent
1983	Kiribati becomes independent
1990	Federated States of Micronesia become independent
1991	Marshall Islands become independent
1994	Palau becomes independent

Pacific Ocean

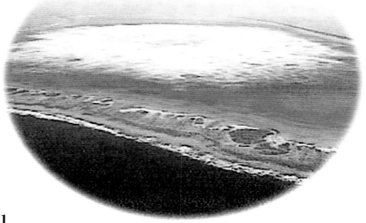

The Pacific Ocean is the largest body of water in the world, with an area of approximately 179,700,000 square kilometres. It is bordered by Asia to the west and the Americas to the east, and is artificially linked to the Atlantic Ocean by the Panama Canal.

The ocean floor is marked by extreme differences in relief. The Marianas Trench plunges to 11,033 metres, the deepest place on the planet, while massive volcanic mountains rise from the ocean floor to form the Hawaiian island chain.

Beneath the Pacific, the cracked crust of the Earth is moving apart at the rate of 20 centimetres a year. Intense activity surrounds the plates, or sections of crust, around the Pacific. This is an area of volcanoes and earthquakes, which in turn may create massive waves called tsunamis. Underwater lava creates chains of volcanic islands, often surrounded by coral reefs.

Ocean currents, including the Californian and Kuroshio, circulate clockwise around the North Pacific, while the Humboldt and East Australian currents move anti-clockwise around the South Pacific. Cycles of warming (an effect known as El Niño) and cooling (La Niña) of waters in the Pacific Ocean have a major effect on global climate, creating severe storms.

▼ **Stone Mysteries**
Giant figures were carved on Easter Island, most easterly of the Polynesian settlements, between the years 1000 and 1600.

▲ **Scorpion fish, Fiji**
Camouflage often conceals venomous spines on these tropical fish.

◄ **Sepik turtle**
This reptile takes its name from the Sepik River in northern Papua New Guinea.

Atlantic Ocean

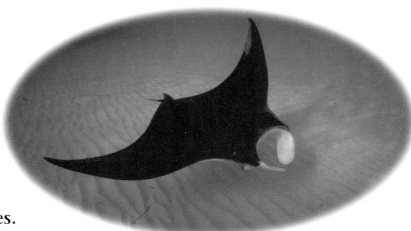

▲ Squid
Atlantic fishermen regard the many species of squids as pests, because they eat herring and mackerel.

The Atlantic is the world's second-largest ocean. It covers a total area of more than 106 million square kilometres. Its average depth is 3,580 metres, but depths of more than 9,000 metres occur in some trenches. The many islands in the Atlantic include Greenland, Iceland and the British Isles.

▲ Manta ray
Manta rays live near the surface in the open sea. They feed on plankton.

▲ Oil rigs
The Atlantic has vast reserves of oil and natural gas, mined by offshore drilling platforms.

The ocean's main feature is the Mid-Atlantic Ridge, an underwater mountain range which stretches from north to south. Running through the ridge is a rift valley, where earthquakes are common and where lava is flowing onto the surface creating new crustal rock. This rift valley is the border between huge plates, which are widening the Atlantic by about 2.5 centimetres a year.

Strong currents flow around the Atlantic Ocean. One famous current, the Gulf Stream, flows from the Gulf of Mexico to northwestern Europe. Its waters bring mild weather to such places as the Norwegian coast. The Atlantic contains important fishing grounds, though some places have been overfished. The Atlantic is also the busiest ocean for shipping.

▼ Trawler
The Atlantic Ocean produces about a third of the world's annual catch of fish and shellfish.

◄ Viking longships
Viking explorers sailed from Europe to Iceland and North America.

▼ Humpback whale
Humpback whales are seen in coastal waters in all the oceans.

Indian Ocean

Green turtle
Hunting has greatly reduced the number of green turtles in the Indian Ocean.

Moray eel
Moray eels live in tropical waters, especially around coral reefs.

The Indian Ocean, the world's third-largest ocean, lies between Africa in the west, Indonesia and Australia to the east, and Antarctica to the south. It covers an area of about 74 million square kilometres. Major islands include Madagascar and Sri Lanka. India and Sri Lanka divide the ocean's northern part into the Bay of Bengal and the Arabian Sea.

The average depth of water is 3,840 metres, but the deepest point is in the Java Trench where depths of more than 7,400 metres have been recorded. Like the Atlantic, the Indian Ocean contains long ocean ridges, where new crustal rock is being formed along central rift valleys. The ridges are also places where earthquakes are common.

The northern part of the Indian Ocean lies in the tropics and water temperatures are high in some areas, especially in the Red Sea and the Persian Gulf. The ocean currents are determined by the winds. North of the equator, the direction of the currents varies every year when wind directions change because of the monsoon.

Sea cucumber
Despite its name, sea cucumbers are animals. In the warm waters of the Indian Ocean, they reach lengths of up to 90 centimetres.

Great white shark
Among the most dangerous sharks, great white sharks have attacked people and even small boats.

Traditional fishing
The Indian Ocean accounts for only five percent of the world's commercial fish and shellfish catch.

237

Polar Lands

Arctic Ocean

▲ **Walrus**

▲ **Harp seals**
Newly-born harp seals were hunted for their valuable soft, white pelts.

The Arctic is the region that lies north of the Arctic Circle. It includes the northern parts of Asia, North America, and Europe, together with most of the world's largest island, Greenland. These land areas enclose the world's smallest ocean, the Arctic Ocean, which covers an area of about 13 million square kilometres. Near its centre is the North Pole.

The Arctic is a bitterly cold region and the Arctic Ocean is largely covered with thick ice throughout the year. This ice blocked early explorers trying to find sea routes around North America and northern Asia. But today, ice-breakers can force their way through the ice. The North Pole can be reached by foot across the ice. The first expedition to reach the North Pole was led by Commander Robert Edwin Peary (1856–1920) of the US Navy in 1909. The first ship to reach the Pole was the nuclear submarine *Nautilus* in 1958.

▲ **Icebound ships**
Ships can become locked in the frozen waters of the Arctic Ocean.

▲ **Inuits**
The Inuit (formerly called Eskimos) live in the Arctic region of North America. In 1999, they took control of a new Canadian territory called Nunavut.

▼ **Drying white fish**
Fishing and hunting seals and whales are traditional activities of Arctic peoples. But today, many people live comfortable lives in permanent settlements.

The average depth of the Arctic Ocean is about 1,120 metres and the greatest depth is 5,450 metres. Whales and various fishes, including cod and halibut, live in the Arctic Ocean. Polar bears live around the ocean and they fish and hunt seals.

The mainland areas in the Arctic contain a large, treeless region called the tundra. Its most common animals are the caribou and reindeer which graze there during the short summer. Other animals include bears, foxes, hares, lemmings and voles. Many birds migrate to the Arctic in summer to breed. Many of them feed on small animals or the insects that swarm in the marshy tundra in summer. Some birds migrate great distances. For example, Arctic terns set out for Antarctica in mid-August. They return to the Arctic in mid-June in the following year.

Arctic peoples include the Inuit of northern Canada and Greenland, the Saami (or Lapps) in northern Scandinavia, and the Samoyeds, Tungus and Yakuts of northern Asia. In the past, most of them lived by fishing and hunting. But many now live in permanent settlements.

FACTS

ARCTIC
Area: 12,000,000 sq km
Average depth: 1,120 m
Greatest known depth: 5,450 m

▶ Inuit dogs
Traditionally, the Inuits used dogs to pull their sleds across the snow. But today, many use motorized snowmobiles to travel over the land.

▲ Sea ice
For most of the year, thick sea ice covers most of the Arctic Ocean.

Yukon

Bering Strait

ALASKA (USA)

Ambarchik

Kolyma

CHUKCHI SEA

Indigirka

Mackenzie

Barrow • Pt. Barrow

EAST SIBERIAN SEA

BEAUFORT SEA

C.Bathurst

New Siberian Islands

R U S S I A

Lena

Banks Island

McClure Strait

C A N A D A

ARCTIC OCEAN

LAPTEV SEA

Victoria Island

Nordvik

Queen Elizabeth Islands

★ North Magnetic Pole

Severnaya Zemlya

Yenisey

★ North Pole

Ellesmere Island

Dikson

Foxe Basin

LINCOLN SEA

Franz Josef Land

Baffin Island

Baffin Bay

Novaya Zemlya

KARA SEA

Ob'

Davis Strait

GREENLAND (DENMARK)

Svalbard (Norway)

BARENTS SEA

Pechora

Godthåb

GREENLAND SEA

North Cape

Murmansk

Denmark Strait

NORWEGIAN SEA

Archangel

ICELAND

• Reykjavik

▲ Polar bear

▼ Bull caribou
Caribou are North American deer. They spend the summer in the Arctic tundra.

▶ Snowy owl

▼ Arctic fox

▼ Icebergs
Icebergs break away from valley glaciers and from Greenland's huge ice sheet. Some drift south into the Atlantic Ocean.

Antarctica

◀ **Wandering albatross**

The Antarctic is the region lying south of the Antarctic Circle. It includes most of the world's fifth-largest continent, Antarctica, which contains the South Pole. Only the tip of the Antarctic Peninsula, jutting out towards South America, lies outside the Antarctic Circle. The waters around Antarctica are sometimes called the Antarctic or Southern Ocean. But most geographers regard these waters as parts of the Pacific, Atlantic and Indian oceans.

▼ **Vital supplies**
Supplies must be flown or shipped in to Antarctica.

Ice covers about 98 percent of Antarctica, although mountain peaks jut through the ice in places. The average thickness of the ice is 2,200 metres, but in places it is 4,800 metres thick. Antarctica is bitterly cold and the world's lowest air temperature, –89.2°C, was recorded at the Russian Vostok Station in 1983. Few plants and animals are found on Antarctica. The best-known creatures are penguins, flightless birds that feed mainly on fish in the waters around the continent.

▼ **Ice cave**
Around the coast of Antarctica, the sea hollows out spectacular caves in the ice.

Early explorers in Antarctica faced great hardship. The first expedition to reach the South Pole, in December 1911, was led by the Norwegian Roald Amundsen (1872–1928). A British expedition led by Robert Falcon Scott (1868–1912) reached the pole five weeks later, but Scott and all his team perished on the way back. Later explorers used aircraft. The first flight over the South Pole was made in 1929 by Richard E Byrd (1888–1957), a US naval officer.

Several countries have claimed parts of Antarctica, but none of the claims is recognized under international law. Many people would like to make the continent a huge international park, a protected wilderness safe from development and pollution.

Today, scientists from several countries work in Antarctica. In the 1980s, scientists discovered that the ozone layer in the stratosphere over Antarctica was being thinned by chemicals called fluorocarbons. The ozone layer is important because it protects our planet from harmful ultraviolet rays. As a result, international action has been taken to reduce the release of fluorocarbons into the atmosphere.

FACTS

ANTARCTICA
Area: *about 14,000,000 sq km*
Population: *none permanent*
Highest point: *Vinson Massif (5,140 m)*

▶ **Trawling near Antarctica**
The waters around Antarctica are rich in marine life, including krill (tiny, shrimp-like creatures), squid, seals, fish and whales.

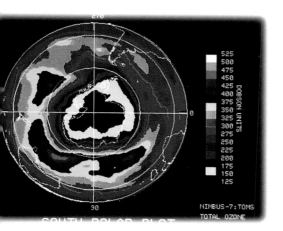

◀ Ozone layer
The ozone layer in the upper atmosphere of the Earth blocks most of the Sun's harmful ultraviolet rays. Since the 1980s, pollution has created 'holes' in the ozone layer over Antarctica.

▲ Penguin parade
Adélie penguins, the most common penguins in Antarctica, build nests of pebbles on the coast.

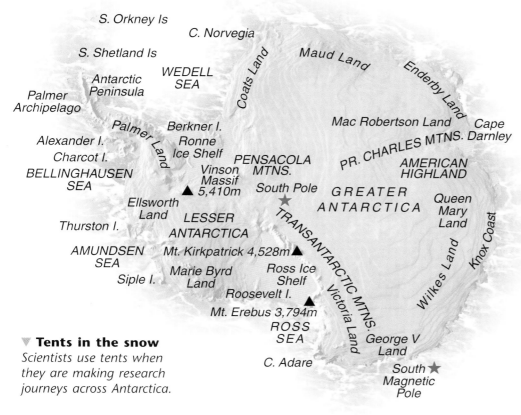

S. Orkney Is
C. Norvegia
S. Shetland Is
Maud Land
Enderby Land
WEDELL SEA
Coats Land
Antarctic Peninsula
Palmer Archipelago
Palmer Land
Mac Robertson Land
Cape Darnley
Alexander I.
Berkner I.
Ronne Ice Shelf
PR. CHARLES MTNS.
Charcot I.
PENSACOLA MTNS.
AMERICAN HIGHLAND
BELLINGHAUSEN SEA
Vinson Massif ▲ 5,410m
South Pole
GREATER ANTARCTICA
Ellsworth Land
LESSER ANTARCTICA
Queen Mary Land
Thurston I.
TRANSANTARCTIC MTNS.
AMUNDSEN SEA
Mt. Kirkpatrick 4,528m ▲
Knox Coast
Siple I.
Marie Byrd Land
Ross Ice Shelf
Wilkes Land
Roosevelt I. ▲
Mt. Erebus 3,794m
Victoria Land
ROSS SEA
George V Land
C. Adare
South Magnetic Pole ★

▲ Roald Amundsen
Amundsen, a Norwegian explorer, led the first expedition to reach the South Pole on December 14, 1911.

▼ Tents in the snow
Scientists use tents when they are making research journeys across Antarctica.

▶ Weather balloons
Weather studies are important, because conditions in Antarctica affect the world's weather.

▼ Whopping whale
Blue whales, the largest animals ever to have lived, feed on the krill in the seas around Antarctica.

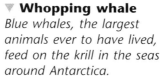

Glossary

Abyssal plain *The deepest part of the ocean, below the continental slope.*

Ammonites *Extinct molluscs whose fossils are common in rocks formed during the Mesozoic era. Like the dinosaurs, they became extinct about 65 million years ago.*

Asteroid *A very small planet in orbit around the Sun. Most are found between the orbits of Mars and Jupiter.*

Atmosphere *The layer of air that surrounds \the Earth.*

Automation *The use of machines to perform tasks that are too repetitive, complex, dangerous or costly for people to undertake*

Biome *A plant and animal community that covers a large area. Biomes include the tundra, coniferous forests, temperate forests, temperate grasslands, deserts, savanna and tropical rainforests.*

Birth and death rates *The number of births and deaths in a year for every 1,000 people.*

Block mountain *A mountain system formed when a block of land is pushed upwards between sets of roughly parallel faults.*

Circumference *The distance around a circle.*

Colony *A territory ruled by a foreign power.*

▲ **Atmosphere**

Comet *A heavenly body usually consisting of a nucleus and tail, composed of frozen gases of ice and dust, that follows a definite path through space.*

▶ **Ammonites**

Communism *A political system where land, industry and all property and goods are controlled by the government.*

Compass *Instrument used to measure direction. Magnetic compasses have needles that point to the magnetic North Pole.* ▲ **Compass**

Condensation *A change of state that occurs when a gas or vapour is turned into a liquid. Invisible water vapour in the air condenses into water droplets when the air is chilled.*

Coniferous forest *Forests of mostly evergreen, cone-bearing trees. Common trees include fir, larch, pine and spruce.*

▲ **Block mountain**

Continent *A large land mass. The world contains seven continents. In order of size, they are Asia, Africa, North America, South America, Antarctica, Europe and Australia.*

Coral *A rock formed from the external skeletons of tiny sea creatures called coral polyps.*

Cumulonimbus cloud *A high, often anvil-shaped thundercloud formed when warm, moist air rises quickly.*

Delta *A flat area at the mouth of a river, formed from sediments deposited by the river.*

Depression *A region of low air pressure, associated with changeable, and often stormy, weather.*

Dialect *A variation of a language used by a group of people in a particular area.*

Diameter *The distance formed by a straight line between the centre of a circle and the circumference.*

▼ **Deciduous trees**

Deciduous trees *Trees that shed their leaves during a certain season. Such deciduous trees as elms and oaks in the mid-latitudes shed their leaves in autumn. In monsoon regions, deciduous trees shed their leaves in the dry season.*

Earthquake *Sudden movements or tremors in the Earth's crust that make the ground shake.*

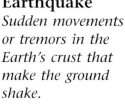

▼ Fossils

Fossil *Evidence of ancient life found in rocks. They include traces of leaves, shells or bones, together with footprints made by animals. Sometimes, whole animals or plants are preserved.*

Fossil fuels *Fuels, such as coal, oil and natural gas, that are formed from the remains of once-living organisms.*

Globe *A model of the Earth consisting of strips of paper showing areas of land and sea pasted on a hollow sphere.*

Habitat *The type of place where a plant or animal lives. Each habitat limits the types and numbers of things that live in it.*

Homo sapiens *The scientific name for human beings, distinguishing them from early human-like creatures, such as Homo erectus and Homo habilis.*

Industrial Revolution *An important event in history involving the use of power-driven machines and the growth of factories. It began in the late 18th century in Britain and, by the mid-19th century, it had spread through much of western Europe and North America. The Industrial Revolution is still continuing in some developing countries.*

International date line *An imaginary line around 180 degrees east and west longitude. When people cross this line from west to east, they gain a day. When they cross from east to west, they lose a day.*

Magma *Molten rock inside the Earth's crust. When it spills on to the surface through volcanoes, it is called lava.*

Environment *The external conditions that influence the growth and development of plants, animals and people.*

Erosion *The process by which natural forces, including weathering, running water, ice, winds and the sea, constantly wear down the land and transport the worn material, only to deposit it elsewhere.*

Ethnic group *A group of people with common characteristics, such as ties of ancestry, culture, language, nationality or religion. These characteristics distinguish them from other people in the same country or society.*

Evaporation *A change of state that occurs when a liquid turns into a gas or vapour. For example, heat evaporates water to create invisible water vapour.*

Fault *A fracture or break in rocks in the Earth's crust along which the rocks have moved. Sudden movements along faults cause earthquakes.*

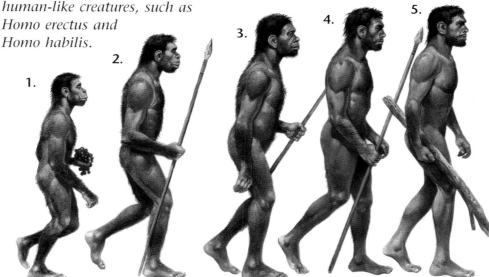

1. Australopithecus
2. Homo habilis
3. Homo erectus
4. Homo sapiens neanderthalensis
5. Homo sapiens sapiens

▼ Earthquake

Hurricane *A severe tropical storm that forms over the oceans north and south of the Equator. When it reaches land, the hurricane often does much damage.*

Hydroelectricity *Electricity that is generated at power stations by the force of water flowing through turbines at dams.*

Magnetic pole *The Earth is like a giant magnet. It has two magnetic poles, north and south just like the ends of a magnet.*

Mangrove swamp *Coastal swamp in tropical regions, in which mangrove trees grow in the salty water. The stilt-like roots sent down from their branches trap silt which often builds up to form new land.*

Map *A representation of the Earth or part of it on a flat surface, such as a sheet of paper.*

Glossary

Meteor *A streak across the sky, also known as a shooting star, caused by a lump of rock burning up as it enters the Earth's atmosphere. Meteoroids that reach the surface are called meteorites.*

Mid-latitudes *The mostly temperate climatic zones between the hot tropics and the cold polar regions.*

Migration *The movement of animals and people from one location to another. People usually migrate to escape from something they do not like, or to go somewhere that seems to be more attractive.*

Minorities *Groups of people who differ in some ways, for example in culture or language, from the main group in a society.*

Natural resources *Materials that are found naturally, such as energy sources, forests, minerals and fertile soils.*

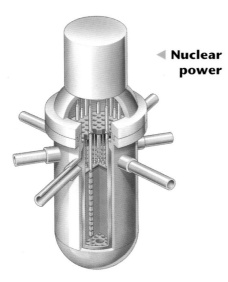

◄ **Nuclear power**

Nuclear power *Energy that is released from a controlled nuclear reaction in a nuclear power station.*

Ozone *An unstable form of the gas oxygen. It is produced when electricity is discharged through air.*

244

Peninsula *A narrow strip of land that juts out into the sea.*

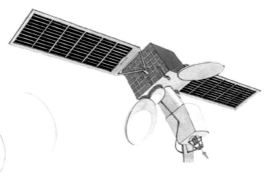

▲ **Satellite**

Pingo *A mound that occurs in polar regions when freezing water beneath the surface pushes up the overlying soil.*

Planet *In our Solar System, a heavenly body that orbits the Sun. Astronomers have identified other distant stars that have planets orbiting around them.*

Plankton *Microscopic plants and animals that float near the surface of oceans. They are the food for many marine creatures.*

Plateau *An upland region with a generally level surface.*

▲ **Stratosphere**

Prairie *A mid-latitude grassland region in North America. It is the equivalent of the pampas in South America and the veld in South Africa.*

Rift valley *A valley formed when a block of land sinks between sets of roughly parallel faults.*

Satellite *In astronomy, a body that revolves about a planet. Artificial satellites are manufactured objects that orbit the Earth or some other heavenly body.*

Smog *A fog that is mixed with smoke. Photochemical smog is caused by the action of sunlight on exhaust gases from cars and factories.*

Soil erosion *The removal of the topsoil as the result of human interference with the land, such as deforestation. Soil erosion occurs quickly on exposed areas, especially on sloping land. By contrast, natural erosion is a much slower process.*

Steppe *A mid-latitude grassland in the Ukraine and extending into central Asia. The term is often used for grasslands that are drier than prairies, with shorter and coarser grasses.*

Stratosphere *The part of the atmosphere above the troposphere, extending to about 130 kilometres above ground level. It contains the ozone layer.*

Volcano *A hole (or vent) in the ground through which lava, steam and gases are ejected. The term volcano is also used for the mountains of ash and lava that form around the vent.*

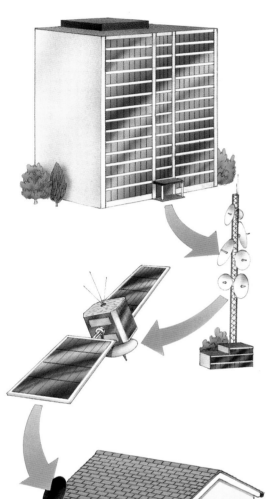

Subsistence farming *The farming of crops in order to support a farmer's family. It is the chief activity in many developing countries and contributes little to their national economies.*

Telecommunications *The communication of information over long distances, by a wide variety of electrical and electronic systems, including telephone, telegraph, radio, television and satellite.*

Temperature *The measurement of the hotness or coldness of a gas, liquid or solid. It is normally measured on the Celsius or Fahrenheit scales.*

Tornado *A small, intense storm, or whirlwind, with wind speeds up to 650 kilometres per hour.*

◀ **Volcano**

▶ **Tornado**

Water vapour *Invisible moisture in the air. It has the properties of a gas, until it condenses into droplets of water.*

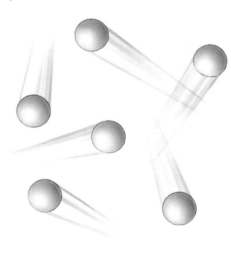

▲ **Molecules of water vapour**

Trade *The buying and selling of goods and services.*

Urban area *A built-up area, such as a town or city and the suburbs around it.*

Wind *The movement of air either across the Earth's surface or within the atmosphere. For example, the jet stream is a powerful wind that blows in the upper atmosphere.*

▲ **Telecommunications**

245

Index of Place Names

250

General Index